A Brief History of Colour Theory

George Pavlidis

A Brief History of Colour Theory

Foundations of Colour Science

 Springer

George Pavlidis
ILSP - Institute for Language and Speech
Processing
ATHENA - Research and Innovation Centre
in Information, Communication
and Knowledge Technologies
University Campus at Kimmeria
Xanthi, Greece

ISBN 978-3-030-87770-5   ISBN 978-3-030-87771-2   (eBook)
https://doi.org/10.1007/978-3-030-87771-2

This Springer imprint is published by the registered company Springer Nature Switzerland AG
The registered company address is: Gewerbestrasse 11, 6330 Cham, Switzerland

Preface

I am Riku Sato and I am recording this in case things go south. Today it is Sunday, 29th of March, 2314, a cold spring morning with around 5° Celsius outside. I am making these notes at home, near Maruyama Park in Sapporo, Hokkaido, north Japan. I am a Biotechnology Engineer for BioSap Co. Ltd., a small company just outside Sapporo, which specialises in agriculture and food manufacturing. I can see from my window the blooming of the cherry trees in Maruyama Park, a sight that has been known to attract attention due to its exceptionally *colourful beauty* for centuries. Actually, this is one of the parts of the metropolis of Hokkaido that remains the same along with some extreme mountainous regions on the island. The rest has been urbanised since the mid-twenty-second century when the majority of the 'civilised' countries turned to the *hedonic society model* after intelligent robots took over almost all the works in every sector. With most of the work being done by intelligent devices, people are now exploring the limits of hedonism in ways I should not discuss further here. Since the world population stabilised to around 9 billion in the twenty-second century due to various reasons that relate to the quality of life decisions and natural evolution factors, this is what happened to the whole world, actually. What is still changing though is life expectancy, which before the nanotechnology enhancements had naturally reached 100 years, in Japan, 110.

Everyday life is interwoven with *the Grid*, the supercomputing network that connects everything on the entire planet. For historical reasons, suffice to say that, as of the early 2200s, nearly a century now, personal supercomputing devices are a commodity, enabled by intelligence-guided design that pushed electronics efficiency to the limits of current physics. I remember from my school years, people used to say that Moore's Law reached its physical limit at that point in history.

About 10 years ago, nanotechnology has, for the first time, been successfully applied to *upgrade* human bodies, leading to what we now call the *transhuman* society. Even though not all people have accepted this change (yet), most have already taken this direction, which practically leads to *immortality*, by enabling the shifting from organic to digital and body switching. Apart from enhancing our senses and abilities, this technology opened the way towards linking with each other and being able to share thoughts and emotions among the all-connected real and artificial consciousnesses on Earth. Speaking about myself, among various enhancements, I have already upgraded my *visual sensation* to be able to experience hyperspectral vision in microscopic and telescopic ranges and still strife to get used to the overwhelming amount of the information and the new perceptions it opened.

For the record, to make the leap towards the new type of society we needed to make another leap. By the end of the twenty-third century, humanity matured to a stable *Type 1 civilisation* (remember Nicolai Kardashev's scale?) and begun steps towards Type 2, to cover the all growing needs for ubiquitous energy.

One of my hobbies that I am developing after I embraced the new human condition is the study of *history*. Now that I gained a much deeper insight into being, all kinds of historical events seem to intrigue me more. Like, for example, when 4 years ago the old Voyager I probe, which begun its space travel in 1977, finally reached the inner part of the Oort Cloud, the region that marks the end of the solar system. This old craft is of historical interest, as it was the first to attempt such a long journey, although many more recent attempts have already surpassed it, thus rendering it a relic of the past space exploration. Apparently, the power systems of the probe have long died out and it will float adrift towards the outer rims of the solar system for some thousands of years. Nevertheless, events like this seem to have a different meaning for me now, which I am able to experience within a broader perspective. This is, actually, also the source of my anxiety and fear, as looking into the past has been marginalised, met with extreme scepticism, virtually banned, although not by law, but by the societal customs since the beginning of the twenty-fourth century. After all, hedonic societies rely on *not looking back* and this is why I am afraid this endeavour of mine might prove rather 'unhealthy'.

It was a week ago that I was looking at the blooming trees in Maruyama Park using my newly enhanced hyperspectral vision and was able to appreciate a new *perception of colour*; I was able to selectively switch between various artificial filters, implanted in my new eyes and visual processing pathways, to block out parts of the light reflected from the trees, or even create various colour composites that revealed interesting aspects of the properties of vegetation, like the health of the trees or their water content and nitrogen concentration. As I facilitated my microscopic vision abilities I was able to zoom into the molecular structure of the leaves and the flowers to experience the real-time workings of plant life. Suddenly, I was overwhelmed with the query of how was it even possible to think that my previous state of vision was sufficient in the first place. This chain of thought led to a need for a further investigation of my previous vision mechanism, but also of the various visual apparatuses that we have devised to mimic that primitive form of vision. I remembered an old technology, called *photography*, some people thought of it also as *art*. Today, with the four-dimensional holographic space-time projections, based on an extension of lattice quantum chromodynamics, that old technology, or art, or whatever it may be called seems like another relic of the past, which is good, right? Well, I thought, why not take a deep breath and some of my time to look into it? After all, everyone has access to everything through the Grid. Although I am not knowledgeable in the domains of the physics of light, optics, imaging chemistry and electronics, as well as human vision biology, my plan was to access the Grid and gain as much insight as possible.

I had to go dark and apply several layers of spoofing and masking to hide my Grid identity for protection, though, which is something I cannot discuss further.

My initial search results were all too unclear, too messed up, too fragmented. I was able to locate excellent sources for the nature of light, for the physiology of the eye, vision and visual perception—which is so different from vision—for the physics of the optics, the photochemistry and the relating reactions, the photoelectric phenomenon and the electronic light sensing and lighting technologies, the history and development of photo-sensing apparatuses. Along these, there were sources for the technical specifications of the human vision, which contradicted each other or agreed to an average, at best. No single source was to be found to provide the full insight into the notions involved. All the sources agreed, though, that any study of vision or visual sensing devices starts with the study of light, moves on the definition of vision, the development of the notion of colour, and finally the analysis of the specifications of the various devices and some almost mystical principles, like that golden ratio rule; all these seemed like constant strife of human intelligence thought the ages, dating back to the antiquity. Among the studies, that of the human visual sensing mechanism, the eye has a prominent position. The study of the visual sensing apparatuses includes in most cases the disintegration of composite devices into their main parts, as those parts have been developed based on totally different principles, although to the same end, to mimic the sensation of vision, but also to correct or alleviate some limitations and naturally occurring defects. In the middle of everything, I found the central notion of colour, its meaning, its scientific and semantic content. Colour became my focus of research and thus the following texts summarise the findings of this investigation, hopefully in an insightful way so that I may have it as a reference for the future, and might be of some utility to others on the Grid.

As I was digging into the Grid, I managed to get my hands on an old twenty-first-century photography apparatus from my good friend Ichika who is a Conservator in the Sapporo Art Museum, among the last such institutions to hold fragments of the human memory, unfortunately with little public interest anymore. This apparatus is called a *photographic camera*, which was supposed to be state of the art at that time. I instantly developed an enthusiasm towards photography as a form of expression and a love for the retro. But what would photography be if it weren't for colour and vision? What is reported in the subsequent texts reflect my ongoing investigation for an in-depth understanding, experimentation and, sometimes, expression. It is my hope that this treatise will aid others in search of such understanding before everything is lost forever.

Sapporo, Japan   Riku Sato
March 2314

Contents

List of Figures

List of Tables

Chapter 1
Introduction

—Παντάπασι δή, ἦν δ᾽ ἐγώ, οἱ
τοιοῦτοι οὐκ ἂν ἄλλο τι
νομίζοιεν τὸ ἀληθὲς ἢ τὰς τῶν
σκευαστῶν σκιάς.

Plato, The Republic

Any presentation of a *theory of colour* presupposes a deep understanding of the nature of colour *per se*. Once understood that *there is no such thing as colour in nature* and that *colour is purely a subjective sensation* fabricated in the brains of the living beings, and developed simply (or, maybe, not so simply) for survival, things get complicated. As colour is not an objective phenomenon in nature, no theory can be developed that could stand if the observer is excluded from the equation. Of course, this insight had not become common sense till the beginning of modern thinking; actually, any layperson would not accept it even today. What is certain to come to mind is the pressing question, *so, how then humans are so certain they see colours?* A scientific response to this question demands two prerequisites, first, an understanding of the nature of light and, second, an understanding of the mechanism of vision and visual perception, as a function of light measurement and interpretation. Scientifically, there is a natural *phenomenon* (light) and a *system* (eyes-brain) to measure and qualify it.

Regarding the *phenomenon*, what is certainly objective is that light is created through energy exchange in the electrons of the atoms and is expressed as electromagnetic radiation, which undergoes some interesting transformations when it travels through space (and media). As radiation, it is characterised primarily by a frequency or a wavelength and an intensity. Particularly, the part of electromagnetic radiation that is visible to the beings on the Earth is called the *visible light* (or light in the visible spectrum) and this is that concerns any study for a theory of colour on this planet. As presented in the section devoted to the nature of light, light on the Earth comes (primarily) from the Sun, and this light has a special composition of various wavelengths. This is a 'composite' light and it 'decomposes' when absorbed by objects or refracted during transmission through the atmosphere, water and objects. This, in effect, results in some objects being able to absorb all of the incident light, thus reflect nothing, other objects (or media) to be able to absorb only a selection of wavelengths and reflect the rest, other objects to 'decompose' the composite light into light rays of single wavelengths that travel in diverging paths, or in other cases, objects cause the scattering of light, to selectively change the paths of various wavelengths.

Regarding the *system* that measures and qualifies light, consisting of the eyes and the brain, it has been a significantly intense endeavour to decode their working through the ages, in some cases hand-in-hand with the research on the light itself. This is a study of the *sense and sensation of light* in humans (and other beings). The understanding of the sense and sensation of vision in humans was among the most interesting philosophical, technical and scientific challenges throughout the ages, with some of the most intelligent and hard-working people involved in this endeavour, many times, with little success. What is clear is that, somehow, humans perceive the light of any particular wavelength (or of different wavelength content) as something totally distinctive and very important for the perception of the world. The main text of this treatise presents a brief history of the foundations of colour science throughout the ages.

In the era of philosophy, prior to the advanced technical civilisation, light and visual perception have been regarded as enormously important concepts that connected with the being and its basic elements. Later this approach was disregarded and scholars turned to new hypotheses and tried to base them on experimental and empirical evidence. For almost 2.5 millennia people strived to find the answers until reaching a minimum of basic understanding, yet not being absolutely certain of the mechanisms involved. The following paragraphs summarise the efforts towards this understanding since ancient times. J.L. Benson (2000) presents an interesting analysis of what he names the *Greek colour theory*, a theory largely influenced by the four-colour system of *white, yellow, red, black* and its connection to the four basic elements in nature, which is based, essentially, on the Greek concepts of polarity and complementation.[1]

There is a large gap in the relevant literature and limited sources can be found in the ages between *the ancients* and *the moderns*, in the middle ages. *The moderns* in this treatise are those giants of thinking that (re-)appeared in the 16th century and reshaped humanity. Their scientific, methodological and analytical approach that was largely influenced by the thinking of the ancients, resulted in our current understanding of visuospatial perception. Nevertheless, in the era of the *middle ages*, there are two persons worth the inclusion in a historical account of the development of the theories of light and colour perception for multiple reasons. They are no other than Abū 'Alī al-Ḥasan ibn al-Ḥasan ibn al-Haytham, or simply Alhazen, and Erasmus Ciołek Vitello (or Witelo).

---

[1] The chapter regarding the colour theory of the ancient Greeks can be found online @ http://scholarworks.umass.edu/art_jbgc/6/. Another important online resource of historical information can be found @ http://www.color-theory-phenomena.nl/08.00.html on Paul Schils' website on *Color Phenomena*.

700 B.C.	600	500	400	300	200
			EMPEDOCLES	ARISTOTLE	ARCHIMEDES
					EUCLID
200 B.C.	100	BIRTH OF CHRIST	100	200	300
				P T O L E M Y	
300	400	500	600	700	800
800	900	1000	1100	1200	1300
			ALHAZEN		R. BACON
					VITELLIO
1300	1400	1500	1600	1700	1800

MAUROLYCUS · JANSEN · DEDOMINIS · GALILEO · SCHEINER · KEPLER · RHEITA · SNELLIUS · DESCARTES · GRIMALDI · BARTHOLIN · HUYGENS · BARROW · MARIOTTE · BOYLE · HOOKE · NEWTON · ROEMER · HAUKSBEE · JURIN · TAYLOR · SMITH · HALL · BRADLEY · BOUGUER · PORTERFIELD · JEAURAT · DOLLOND · L. EULER · SIMPSON · CLAIRAUT · DALEMBERT · KLINGENSTIERNA · LAMBERT · DUTOUR · BOSCOVICH · PRIESTLEY · RAMSDEN · RIN

Fig. 1.1 The timeline of the development of Optics according to Thomas Young (1807)

The modern era of colour theory is considered to begin with Issac Newton who was deeply concerned with the understanding of the nature of light and colour. Although the ancients and Alhazen and Vitello (among others not listed in this brief account) suggested the modern way of research, it is Newton's approach that marks the beginning of a totally new era in scientific investigation. Although many researchers disagreed with his views about the nature of light and colour, a new road was paved towards even greater challenges. The period after his contributions, several theories were developed but those of Young, Maxwell and Helmholtz had the most significant impact and outlasted any other. During the mid-19th century, the three scientists, either independently or by adopting, rejecting and modifying (or possibly even copying) each other's theories (Heesen, 2015; Kremer, 1993; Sherman, 1981), came up with a theory that was pervasive during the subsequent centuries and was criticised only by a few researchers that proposed alternative models of colour theory. The presentation of the developments in the colour theory of the modern era concludes at the end of the 20th century when all theories converged to a dual nature of light, although to an incomplete theory of vision, thus it is mainly concerned with the scientists that built the basis for a more concrete understanding.

Thomas Young, in 1807 published his paper *On the History of Optics*, in which he summarised the contributions of scholars in the domain since antiquity (Young, 1807c). He created an illustration showing the timeline of the history of optics which is shown in Fig. 1.1, adapted from that paper.

Chapter 2
Colour Theory in the Ancient Times

*—Περὶ μὲν οὖν τοῦ ἄνευ
φωτὸς μὴ ὁρᾶν εἴρηται ἐν
ἄλλοις· ἀλλ᾽ εἴτε φῶς εἴτ᾽ ἀήρ
ἐστι τὸ μεταξὺ τοῦ ὁρωμένου
καὶ τοῦ ὄμματος, ἡ διὰ τούτου
κίνησίς ἐστιν ἡ ποιοῦσα τὸ
ὁρᾶν.*

Aristotle, Sense and the sensible, 438b

2.1 Introduction

In Homer's Odyssey (Οδύσσεια του Ομήρου c.8th century BCE), rhapsody 19, verses 172–173 one may read[1]

Κρήτη τις γαῖ᾽ ἔστι, μέσῳ ἐνὶ οἴνοπι πόντῳ,
καλὴ καὶ πίειρα, περίρρυτος·

which was (poetically) translated by N. Kazantzakis and I. Kakridis into modern Greek as[2]

Μια χώρα, η Κρήτη, μέσα βρίσκεται στο πέλαο το κρασάτο,
περίσσια πλούσια, θαλασσόζωστη, πανώρια

verbatim ac litteratim these verses could be translated to

---

[1] For the original Greek text of Homer's Odyssey in rhapsody 19 see the online resource @ https://www.greek-language.gr/digitalResources/ancient_greek/library/browse.html?text_id=133&page=144.

[2] This text can be accessed online @ https://archive.org/details/u_20201210/page/n259/mode/2up, pg.265 (in the PDF file provided).

5
G. Pavlidis, *A Brief History of Colour Theory*,
https://doi.org/10.1007/978-3-030-87771-2_2

> There is a land called Crete, in the middle of an open
> sea that looks like wine
> beautiful and fertile, surrounded by water;

Apparently, under almost no circumstances one would, today, be ready to accept a body of water (the open sea in this case) resembling wine in appearance. Even under very specific conditions, like during dusk or dawn, or even under particular weather conditions, it is not something easily imagined. Nevertheless, Homer, one of humanity's founders of literature, supposedly blind, was using a mixture of the Ionic and Aeolic dialects of the ancient Greek language of his era (the Epic Greek, as usually referenced) to describe the open sea (the Aegean see) as looking like wine. Many scholars have been trying to tackle this particular (and other such differentiations to modern terminology) and suggested that Homer was most probably referring to a hue of dark colour *mauve*,[3] a particular dark purple hue that might actually appear both as a colour for wine and as a colour for a deep sea under certain illumination and weather conditions. Many differences in the usage of language to describe colours can be identified to date. The names and the meanings of colours vary through the ages and the peoples. Such differentiation can be traced as far as the 19th century, with the 20th century being a landmark in applying standardisation and resolving most ambiguities regarding colour.

Figure 2.1 shows a photo of the masterpiece fresco *The School of Athens* by Raphael (Raffaello Sanzio da Urbino) that was painted between 1509–1511 in the Apostolic Palace in the Vatican. At the centre of the fresco, Plato walks alongside Aristotle, whereas several other well-known thinkers discuss various subjects (of which only some are Athenians regardless of the fresco title). It has been suggested that in this fresco many known philosophers are depicted, including Plato and Aristotle, Anaximander, Pythagoras and Archimedes, Socrates, Heraclitus, Zeno and Parmenides. It has long been discussed that there is very strong symbolism in the representation and posture of the body of each of the philosophers, which is particularly interesting for the case of the central figures, Plato and Aristotle. Plato, who believed in an absolute world of ideas, is shown holding his work *Timeaus* and pointing to the heavens (theory of forms, the timeless ideal), whereas Aristotle, who believed in observation, trust of the senses and reasoning, is shown holding his work *Nicomachean Ethics* and has a more relaxed open hand towards the ground (the present, life, grounding). This section of this treatise discusses the contribution of the philosophers of ancient Greece to the efforts towards a theory of colour.

---

[3] Interestingly, *mauve* sounds so close to *mavì*, from the Arabic مَاوِيّ (māwiyy) which means *watery*.

Fig. 2.1 The School Of Athens by Raffaello Sanzio Da Urbino

2.2 Alcmaeon of Croton

Among the first pieces of evidence of ancient Greeks trying to understand visuospatial perception can be identified in the work of *Alcmaeon* (Αλχμαίων ο Κροτωνιάτης), who lived in Croton, in Magna Graecia (Southern Italy), during the end of the 6th and the beginning of the 5th century BCE. What is known about him and relates to the topics in this treatise come from Theophrastus, Chalcidius, Aetius and Stobaeus. Apparently, Alcmaeon was among the first to undertake dissection and to demonstrate the nature of the eyes. He proposed that the eyes are connected with water, although fire is also part of their nature, and that sight depends on brilliance and transparency of reflection. He was also among those who proposed that the brain is the governing centre to which all the senses are connected.

He is said to have been a Pythagorean, but this could not be verified. Alcmaeon influenced other great minds, like Empedocles and Democritus (Wachtler, 1896; Codellas, 1932; Diels & Kranz, 1903). As listed in Diels and Kranz (1903)[4], in Theophrastus' *On Sensation* there is a characteristic passage marking that the eyes see via the water, and it is obvious that the water in the eyes encloses fire, for when the eyes are struck sparks are created; the eyes see through the gleaming and the transparent when light is reflected, and the greater the purity the better; all the senses are then concentrated to the brain.

---

[4] The 1903 edition can be found online at https://archive.org/details/diefragmenteder00krangoog/.

ὀφθαλμοὺς δὲ ὁρᾶν διὰ τοῦ πέριξ ὕδατος· ὅτι δ᾽ ἔχει πῦρ
δῆλον εἶναι, πληγέντος γὰρ ἐκλάμπειν. ὁρᾶν δὲ τῷ στίλ-
βοντι καὶ τῷ διαφανεῖ, ὅταν ἀντιφαίνῃ, καὶ ὅσον ἂν κα-
θαρώτερον ἢ μᾶλλον. ἁπάσας δὲ τὰς αἰσθήσεις συνηρτῆσ-
θαι πως πρὸς τὸν ἐγκέφαλον.

2.3 Parmenides

Parmenides (Παρμενίδης ο Ελεάτης, c.540-470 BCE) was born in Elea of Magna
Graecia (Southeastern Italy) was a Greek philosopher, of whom a single work is
known, a poem usually referenced as *On Nature* (typical title for many ancient
philosophers, although unverified). What is known of this poem comes from the
works of Plato, Sextus Empeiricus, Proklos, and Simplicius. In this extraordinary
and far-reaching work of a genius, Parmenides presented two views of reality, the
objective world of absolute truths (ἀλήθεια-*on truth* or *on the intelligible*) and the
subjective world of human sensation (δόξα-*on opinion* or *on the perceptible*). In the
first view, Parmenides presents his theory of *Being* as a timeless and dimensionless
entity that can only exist and be absolutely true and stationary, whereas, in the second
view, he lays out his cosmology of the perceived world, in which he claims that it is
an illusion, always in motion and false.

Apparently, Parmenides recognised the limitations of human perception and differentiated the world of philosophy (truth) to the world of life (opinion) (Davidson,
1869; Diels & Kranz, 1903). In this view, visual perception is basically an illusion
and there is no such thing as colour (Ierodiakonou, 2005). Parmenides states that
since the Being is stationary, absolute and Real, that which humans perceive and
name as true and ever-changing (like colour or brilliance) is false (Davidson, 1869
(II); Diels & Kranz, 1903 (B 8.38–41)),

τῷ πάντ᾽ ὄνομ(α) ἔσται,
ὅσσα βροττοὶ κατέθεντο πεποιθότες εἶναι ἀληθῆ,
γίγνεσθαι τε καὶ ὄλλυσθαι, εἶναι τε καὶ οὐχὶ,
καὶ τόπον ἀλλάσσειν διά τε χρόα φανὸν ἀμείβειν.

2.4 Empedocles

Empedocles (Ἐμπεδοχλής ο Ακραγαντινός, c.495-430 BCE) was among the pioneers for a theory of light, vision and perception. He lived in Acragas (Agrigentum), in Magna Graecia (Southern Sicily). His theory can be found in various sources, including Theophrastus, Plato and Aristotle. References for his work and life can be found in Diels and Kranz (1903), where parts of his most important work *On Nature* is included. This work was written as a poem of hexameter verse, counting about two thousand verses (Ierodiakonou, 2005). Apparently, Empedocles supported a theory of a Universe consisting of four basic elements (roots), from which two have a colour. These roots are fire, water, earth and air, of which fire is white and water is black. He seemed to have argued that all colours are the result of combinations of white and black, not as mixtures, but as compact arrangements of tiny portions of them (like infinitesimal dots of black and white). According to Flavius Aetius (c.391–454), Empedocles was supporting *a theory of four primary colours, white, black, red and yellow*, but this cannot be actually confirmed from the sources in Diels and Kranz (1903), and Ierodiakonou (2005) rightfully rejects it based also on other saved context information.

Empedocles supported an interesting unified theory of perception, by which each sense organ is 'tuned' to external *effluences* by matching the 'size' of the stimulus with the size (and possibly the nature) of the 'pores' (input pathways) of the sense organ. Particularly for the eye, he suggested that it consists of a kind of black-white grating, or a type of regular lattice, to filter and quantify the optical effluences (Sedley, 1992; Ierodiakonou, 2005; Diels & Kranz, 1903). Characteristic is the passage in Aristotle's Metaphysics, Book A, regarding Empedocles' theory of the four basic elements, which is based on including earth into the previous systems of air, water and fire, proposed by Anaximenes, Diogenes, Hippasus and Heraclitus.

Ἀναξιμένης δὲ ἀέρα καὶ Διογένης πρότερον ὕδατος καὶ μάλιστ᾽ ἀρχὴν τιθέασι τῶν ἁπλῶν σωμάτων, Ἵππασος δὲ πῦρ ὁ Μεταποντῖνος καὶ Ἡράκλειτος ὁ Ἐφέσιος, Ἐμπεδοκλῆς δὲ τὰ τέτταρα, πρὸς τοῖς εἰρημένοις γῆ προστιθεὶς τέταρτον, (ταῦτα γὰρ ἀεὶ διαμένειν καὶ οὐ γίγνεσθαι ἀλλ᾽ ἢ πλήθει καὶ ὀλιγότητι, συγκρινόμενα καὶ διακρινόμενα εἰς ἕν τε καὶ ἐξ ἑνός)·

Again in Theophrastus' *On Sensation*, one may read, in a part of a long explanation on the way the sense of vision works, that Empedocles supported that peculiar grating (pore) theory of senses, in which vision is based on white (fire) and black (water) sensing pores next to each other.

Ἐμπερδοκλῆς δὲ περὶ ἁπασῶν ὁμοίως λέγει καὶ φησι τῷ ἐν-
αρμόττειν εἰς τοὺς πόρους τοὺς ἑκάστης αἰσθάνεσθαι· διὸ
καὶ οὐ δύνασθαι, τὰ ἀλλήλων κρίνειν, ὅτι τῶν μὲν εὐρύτεροί
πως, τῶν δὲ στενώτεροι τυγχάνουσιν οἱ πόροι πρὸς τὸ
αἰσθητόν, ὡς τὰ μὲν οὐχ ἁπτόμενα διευτονεῖν τὰ δ᾽ ὅλως
εἰσθελθεῖν οὐ δύνασθαι. πειρᾶται δὲ καὶ τὴν ὄψιν λέγειν,
ποία τίς ἐστι· καὶ φησὶ τὸ μὲν ἐντὸς αὐτῆς εἶναι πῦρ, τὸ δὲ
περὶ αὐτὸ γῆν καὶ ἀέρα δι᾽ ὧν διιέναι λεπτὸν ὂν καθάπερ τὸ
ἐν τοῖς λαμπτῆρσι φῶς. τοὺς δὲ πόρους ἐναλλὰξ κεῖσθαι
τοῦ τε πυρὸς καὶ τοῦ ὕδατος, ὧν τοῖς μὲν τοῦ πυρὸς τὰ
λευκά, τοῖς δὲ τοῦ ὕδατος τὰ μέλανα γνωρίζειν· ἐναρμότ-
τειν γὰρ ἑκατέροις ἑκάτερα. φέρεσθαι δὲ τὰ χρώματα πρὸς
τὴν ὄψιν διὰ τὴν ἀπορροήν.

As Clegg (2002) points out, Empedocles poetically attributes vision to the goddess of love, Aphrodite, who kindled the fire of the eye and confined it with tissues in the eyeball.

2.5 Protagoras

Protagoras (Πρωταγόρας ο Αβδηρίτης c.490-420 BCE, from Abdera, Thrace, Northern Greece) is mentioned by many other philosophers (especially Plato) as the inventor of the profession of the sophist (Diels & Kranz, 1903). Although he is presented as a controversial individual that was mainly concerned with the matters connecting to virtue and politics and he is not directly related with the topics of immediate interest in this treatise, Protagoras is worth mentioning due to his philosophy of relativism, and a famous quote that is of particular and universal interest. In Plato's *Theaetetus*, Socrates quotes Protagoras in saying[5]

Σωκράτης: Κινδινεύσεις μέντοι λόγον οὐ φαῦλον εἰρηκέναι
περὶ ἐπιστήμης, ἀλλ᾽ ὃν ἔλεγε καὶ Πρωτοαγόρας. τρόπον
δὲ τινα ἄλλον εἴρηκε τὰ αὐτὰ ταῦτα. φησὶ γάρ που πάντων
χρημάτων μέτρον ἄνθρωπον εἶναι, τῶν μὲν ὄντων ὡς ἔτσι,
τῶν δὲ μὴ ὄντων ὡς οὐκ ἔστιν.

The passage (somewhat freely) translates to the significant message that *man is the measure of all things, of those that exist that they do, of those that do not exist that they do not.* Further in the same text ([154a]),

---

[5] The passage of the original text is found online @ http://www.hellenicaworld.com/Greece/ Literature/Platon/gr/Theaititos.html, Theaetetus[152a]. The passage can also be found in Diels and Kranz (1903).

Σωκράτης: τί δέ; ἄλλῳ ἀνθρώπῳ ἆρ' ὅμοιον καὶ σοὶ φαίνε-
ται ὁτιοῦν; ἔχεις τοῦτο ἰσχυρῶς, ἢ πολὺ μᾶλλον ὅτι οὐδὲ
σοὶ αὐτῷ ταὐτὸν διὰ τὸ μηδέποτε ὁμοίως αὐτὸν σεαυτῷ
ἔχειν;

Socrates is saying, in a form of a series of questions, that there is not a thing that looks the same to anyone, not even to the same person, as the same person constantly changes. It seems that Protagoras meant that the sensation triggered by an object to a person is true to a person, whatever 'true' may mean to that particular person. Plato, explains thus, that the meaning of Protagoras' phrase and idea is that there is *no objective truth*, and whatever one believes to be the truth, it is true. Apparently, Plato who believed in a world of ideas (ideals) cannot accept such a position and this is why he contradicted Protagoras, but thanks to this contradiction we know of Protagoras' remarkable idea of relativism, which, in part, is entirely true for the senses and sensations that are the topic in this treatise.

2.6 Democritus

Democritus (Δημόκριτος ο Αβδηρίτης, c.460-370 BCE, from Abdera, Thrace, Northern Greece), the pioneer of physics, used his atomic theory to explain the appearance of colours based on the characteristics of their constituent atoms. In a way, Democritus linked colours with quantities of energy and their shapes, roughness and density, providing a picture, in which colours seem to be similar to energy signatures of the objects themselves. As shown in Diels and Kranz (1903) and later in Enriques and Mazziotti (1948) (translated to Greek in Enriques & Mazziotti, 1982), Theophrastus presented Democritus' theory and was extremely critical of it. He summarised Democritus' theory of vision as a process, by which vision is the result of the impressions (literary, imposed by pressure) of the air between the objects and the eyes. The air assumes the shape of the objects as a result of pressure between the objects and the eyes, and images are formed within the eyes due to the liquid nature of the eyes. According to Theophrastus, Democritus proposed *four primary colours (white, black, red and green)*,[6] of which white and black seem to be more fundamental since white is the colour created by small and smooth particles and their regular arrangement, and black is somehow the opposite (rough and irregularly arranged particles). In a way, this theory is close to not accepting the existence of colours as other thinkers proposed, and this is pointed out by Aristotle and Aetius (Diels & Kranz, 1903). Democritus realised the incompleteness of human perception and the purely subjective nature of the sensations; he held that we cannot judge the absolute truth

---

[6] Democritus used the Greek term χλωρόν (chloron, like in chlorophyll), which can be translated to *light green* but also *tender*.

through sensual impressions, since these are different for each individual even when receiving the same senses, as characteristically expressed in Aristotle's *Metaphysics* (Aristotle, 1993; Tredennick, 1933).[7]

> ἔτι δὲ καὶ πολλοῖς τῶν ἄλλων ζῴων τἀναντία [περὶ τῶν
> αὐτῶν] φαίνεσθαι καὶ ἡμῖν, καὶ αὐτῷ δὲ ἑκάστῳ πρὸς αὑτὸν
> οὐ ταὐτὰ κατὰ τὴν αἴσθησιν ἀεὶ δοκεῖν. ποῖα οὖν τούτων
> ἀληθῆ ἢ ψευδῆ, ἄδηλον· οὐθὲν γὰρ μᾶλλον τάδε ἢ τάδε
> ἀληθῆ, ἀλλ' ὁμοίως. διὸ Δημόκριτός γέ φησιν ἤτοι οὐθὲν
> εἶναι ἀληθὲς ἢ ἡμῖν γ' ἄδηλον.

The passage strongly suggests that either there is no truth or objective truth is not accessible.

2.7   Plato

Plato (Πλάτων ο Αθηναίος 427-347 BCE, from Athens), in his most renowned work *The Republic* (Plato, 2014; Shorey, 1935a; 1935b)–in which he outlined the ideal State in all dimensions, interweaving philosophy, metaphysics, current knowledge, reasoning and beliefs–included the well known *allegory of the cave* (Book VII) to illustrate the difference made by education, yet also to highlight the limitations of human perception and to distinguish a subjective sensation to an objective fact.[8]

> Παντάπασι δή, ἦν δ' ἐγώ, οἱ τοιοῦτοι οὐκ ἂν ἄλλο τι νομί-
> ζοιεν τὸ ἀληθὲς ἢ τὰς τῶν σκευαστῶν σκιάς.

Nothing else would they believe to be true but the shadows, Plato states for the chained prisoners in the cave, those who are more than happy with their perception of the world, which Plato makes sure to inform it is false, from the beginning and in principle. Figure 2.2[9] is a photo of a very interesting depiction of Plato's allegory of

---

[7] The original text with English translation can be found online at https://archive.org/details/in.ernet. dli.2015.185284/mode/2up. The referenced passage and its translation can be found on pp.184–185.

[8] Plato, at the same time, offered one of the very first examples of a *thought experiment*. The original text with an English translation can be found online in two Volumes at https://archive.org/details/republicshorey01platuoft/mode/2up, https://archive.org/details/republicshorey02platuoft/mode/2up.

[9] This figure is a cropped and adjusted reproduction based on the Public Domain image from Wikimedia, found at https://upload.wikimedia.org/wikipedia/commons/b/b1/Platon_Cave _Sanraedam_1604.jpg.

Fig. 2.2 Jan Saenredam's engraving of Plato's Allegory of the Cave

the cave, created in 1604 as an engraving by the Dutch printmaker Jan Saenredam, after a painting by Cornelis Corneliszoon van Haarlem. In his subsequent work, *Timaeus* (Plato, 2008; Archer-Hind, 1888), Plato touched upon more specific matters and laid out his theory about senses, of which he recognised *sight* as being the most important.[10]

> Ὅταν οὖν μεθημερινὸν ᾖ φῶς περὶ τὸ τῆς ὄψεως ῥεῦμα, τότε ἐκπῖπτον ὅμοιον προς ὅμοιον, συμπαγὲς γενόμενον, ἓν σῶμα οἰκειωθὲν συνέστη κατὰ τὴν τῶν ὀμμάτων εὐθυωρίαν, ὅπηπερ ἂν ἀντερείδῃ τὸ προσπῖπτον ἔνδοθεν πρὸς ὃ τῶν ἔξω συνέπεσεν. ὁμοιοπαθὲς δὴ δι᾽ ὁμοιότητα πᾶν γενόμενον, ὅτου τε ἂν αὐτό ποτέ ἐφάπτηται καὶ ὃ ἂν ἄλλο ἐκείνου, τούτων τὰς κινήσεις διαδιδὸν εἰς ἅπαν τὸ σῶμα μέχρι τῆς ψυχῆς αἴσθησιν παρέσχετο ταύτην ᾗ δὴ ὁρᾶν φαμεν.

According to this statement, vision is the result of the coalescence of the stream of fire (light) due to the daylight (the sun) and the stream of fire coming out from the

---

[10] An English translation can be found at https://genius.com/Plato-timaeus-full-text-annotated. Another source of the original text and an English translation can be found at https://archive.org/details/timaeusofplato00platiala/mode/2up.

eyes (the stream or ray of sight). This coalescence produces the sight of objects at the exact location the rays cross each other. Plato proceeded in defining the basic human sensations and described the fourth class of sensation (after taste, smell and hearing), as one that demands fine distinctions, since it appears in many variations, which he called *colours*, as shown in the original text that follows.

> Τέταρτον δὴ λοιπὸν ἔτι γένος ἡμῖν αἰσθητικόν, ὃ διελέσ-
> θαι δεῖ συχνὰ ἐν ἑαυτῷ ποικίλματα κεκτημένον, ἃ σύμπαντα
> μὲν χρόας ἐκαλέσαμεν, φλόγα τῶν σωμάτων ἑκάστων ἀπορ-
> ρέουσαν, ὄψει σύμμετρα μόρια ἔχουσαν πρὸς αἴσθησιν...τὰ
> φερόμενα ἀπὸ τῶν ἄλλων μόρια ἐμπίπτοντά τε εἰς τὴν ὄψιν
> τὰ μὲν ἐλάττω, τὰ δὲ μείζω, τὰ δ᾽ ἴσα τοῖς αὐτῆς τῆς ὄψεως
> μέρεσιν εἶναι· τὰ μὲν οὖν ἴσα ἀναίσθητα, ἃ δὴ καὶ διαφανῆ
> λέγομεν...

Plato states that *colour* is the fire that flows from all bodies and becomes perceptible as it consists of particles that are consistent (compatible) with the ray of sight (which he already defined as another stream of fire coming from the eyes). Apparently, the particles that flow from all bodies can either be equal or unequal to the particles of the ray of sight. The equal particles are imperceptible and are called *transparent*. The larger impose a contraction and the smaller a dilation of the ray of sight, like the cold and hot objects in touch. To name these actions, *white* is defined as the perceived quality that connects with the dilation and *black* connects with the contraction of the ray of sight. Plato followed this line of reasoning to define the mixing process by which to produce more and more colours.

2.8 Aristotle

Aristotle (Ἀριστοτέλης ο Σταγειρίτης, 384-322 BCE, from Stagira, Northern Greece) in his treatises *On the soul* (Aristotle, 2001; Thomas, 1808; Hett, 1935) and *The sense and the sensible* (Aristotle, 2004; Thomas, 1808; Hett, 1935) (among other treatises) was also concerned with the understanding and the definition of visual perception.

Not only he presented his worldview, but he confronted the ideas of some of the greatest minds of his time, like Heraclitus, Empedocles, Democritus, Pythagoras and Plato, whose views he found to be incomplete. In his treatise *On the soul*–which is more like biology in modern terminology–in Book II, Chap. 7, he defines light and colour[11] as follows.

> Οὗ μὲν οὖν ἐστιν ἡ ὄψις, τοῦτ' ἐστὶν ὁρατόν, ὁρατὸν δ' ἐστὶ χρῶμά τε καὶ ὃ λόγῳ μὲν ἔστιν εἰπεῖν, ἀνώνυμον ... τὸ γὰρ ὁρατόν ἐστι χρῶμα, τοῦτο δ' ἐστὶ τὸ ἐπὶ τοῦ καθ' αὐτὸ ὁρατοῦ· καθ' αὐτὸ δὲ οὐ τῷ λόγῳ, ἀλλ' ὅτι ἐν ἑαυτῷ ἔχει τὸ αἴτιον τοῦ εἶναι ὁρατόν. πᾶν δὲ χρῶμα κινητικόν ἐστι τοῦ κατ' ἐνέργειαν διαφανοῦς, καὶ τοῦτ' ἐστὶν αὐτοῦ ἡ φύσις· διόπερ οὐχ ὁρατὸν ἄνευ φωτός, ἀλλὰ πᾶν τὸ ἑκάσ-του χρῶμα ἐν φωτὶ ὁρᾶται. διὸ περὶ φωτὸς πρῶτον λεκτέον τί ἐστιν. ἔστι δή τι διαφανές. διαφανὲς δὲ λέγω ὃ ἔστι μὲν ὁρατόν, οὐ καθ' αὐτὸ δὲ ὁρατὸν ὡς ἁπλῶς εἰπεῖν, ἀλλὰ δι' ἀλλότριον χρῶμα ... φῶς δέ ἐστιν ἡ τούτου ἐνέργεια, τοῦ διαφανοῦς ᾗ διαφανές. δυνάμει δέ, ἐν ᾧ τοῦτ' ἐστί, καὶ τὸ σκότος. τὸ δὲ φῶς οἷον χρῶμά ἐστι τοῦ διαφανοῦς, ὅταν ᾖ ἐντελεχείᾳ διαφανὲς ὑπὸ πυρὸς ἢ τοιούτου οἷον τὸ ἄνω σῶμα· καὶ γὰρ τούτῳ τι ὑπάρχει ἓν καὶ ταὐτόν.

Aristotle states that what is visible is colour and a certain type of entity that can be described in words but has no single name. Whatever is visible is colour. Colour is what lies upon whatever is inherently visible; 'inherently' meaning not that it is visible due to its substance but due to containing the cause of being visible. Every colour is able to set in motion the transparent. That is exactly why it is not visible without the presence of light. Apparently, there is something that is transparent, meaning what is not visible in itself, but rather owing its visibility to the colour of something else. Light is the consistency of something transparent being transparent.

In *The sense and the sensible*, Aristotle outlined his theory of visual sensation and perception. He confronted Empedocles' idea that light emanates from the eye and thus vision is created, as he realised that Empedocles' reasoning was flawed by allowing also emanations (effluences) from the objects.[12]

---

[11] The passage in Greek was adopted *verbatim ac litteratim* from https://goo.gl/QGp7QN.

[12] The passages in Greek that appear in the next paragraphs are adopted from https://goo.gl/i8Pa8G.

Ἐμπεδοκλῆς δ' ἔοικε νομίζοντι ὁτὲ μὲν ἐξιόντος τοῦ φωτός,
ὥσπερ εἴρηται πρότερον, βλέπειν· λέγει γοῦν οὕτως· ὡς δ'
ὅτε τις πρόοδον νοέων ὡπλίσσατο λύχνον χειμερίην διὰ
νύκατα, πυρὸς σέλας αἰθομένοιο, ἄψας παντοίων ἀνέμων
λαμπτῆρας ἀμοργούς, οἵ τ' ἀνέμων μὲν πνεῦμα διασκιδ-
νᾶσιν ἀέντων, πῦρ δ' ἔξω διαθρῷσκον, ὅσον ταναώτερον
ἦεν, λάμπεσκεν κατὰ βηλὸν ἀτειρέσιν ἀκτίνεσσιν· ὡς δὲ τότ'
ἐν μήνιγξιν ἐεργμένον ὠγύγιον πῦρ λεπτῇσιν τ' ὀθόνῃσι
λοχεύσατο κύκλοπα κούρην· αἳ χοάνῃσι δίαντα τετρήατο
θεσπεσίῃσιν· αἳ δ' ὕδατος μὲν βένθος ἀπέστεγον ἀμφιναέν-
τος, πῦρ δ' ἔξω δίεσχον, ὅσον ταναώτερον ἦεν. ὁτὲ μὲν
οὖν οὕτως ὁρᾶν φησίν, ὁτὲ δὲ ταῖς ἀπορροίαις ταῖς ἀπὸ τῶν
ὁρωμένων.

He went on to face Democritus, who held that the explanation of seeing is in the
study of mirrors. Aristotle almost made fun of Democritus stating that it should have
occurred to him to ask why all objects on which images are reflected cannot see and
only the eyes see, in a passage that is rather interesting to read.

Δημόκριτος δ' ὅτι μὲν ὕδωρ εἶναί φησι, λέγει καλῶς, ὅτι
δ' οἴεται τὸ ὁρᾶν εἶναι τὴν ἔμφασιν, οὐ καλῶς· τοῦτο μὲν
γὰρ συμβαίνει ὅτι τὸ ὄμμα λεῖον, καὶ ἔστιν οὐκ ἐν ἐκείνῳ
ἀλλ' ἐν τῷ ὁρῶντι· ἀνάκλασις γὰρ τὸ πάθος, ἀλλὰ καθόλου
περὶ τῶν ἐμφαινομένων καὶ ἀνακλάσεως οὐδέν πω δῆλον ἦν,
ὡς ἔοικεν. ἄτοπον δὲ καὶ τὸ μὴ ἐπελθεῖν αὐτῷ ἀπορῆσαι
διὰ τί ὁ ὀφθαλμὸς ὁρᾷ μόνον, τῶν δ' ἄλλων οὐδὲν ἐν οἷς
ἐμφαίνεται τὰ εἴδωλα.

He agreed with Democritus, though, in that the eye is composed of water, and supported it by evidence and reasoning. Aristotle was extremely critical to those believing that rays of light emanate from the eye itself to intersect other light rays and
create vision. He found it meaningless to accept that something is issuing from the
eye, extending itself to the stars (as he said) or to whatever rays from objects coalesce
with its path to create vision.

ἄλογον δὲ ὅλως τὸ ἐξιόντι τινὶ τὴν ὄψιν ὁρᾶν, καὶ ἀποτεί-
νεσθαι μέχρι τῶν ἄστρων, ἢ μέχρι τινὸς ἐξιοῦσαν συμφύεσθαι, καθάπερ λέγουσί τινες. τούτου μὲν γὰρ βέλτιον τὸ ἐν
τῇ ἀρχῇ συμφύεσθαι τοῦ ὄμματος. ἀλλὰ καὶ τοῦτο εὔηθες·
τό τε γὰρ συμφύεσθαι τί ἐστι φωτὶ πρὸς φῶς, ἢ πῶς οἷόν
θ᾽ ὑπάρχειν (οὐ γὰρ τῷ τυχόντι συμφύεται τὸ τυχόν), τό τ᾽
ἐντὸς τῷ ἐκτὸς πῶς; ἡ γὰρ μῆνιγξ μεταξύ ἐστιν.

He went on asking himself (in a rhetoric manner) *what the meaning is of a coalescence of light with light and how this is possible.* Aristotle concluded Chapter 2 of *Sense and the sensible* by stating that without light, vision is impossible and that vision is caused by a process through a medium between the eye and the objects, which has to be transparent; the same reasoning demands that inside the eye be a translucent substance, water. He further realised that the *soul*, or its perceptive part, is not at the external surface of the eye, but somewhere within.

περὶ μὲν οὖν τοῦ ἄνευ φωτὸς μὴ ὁρᾶν εἴρηται ἐν ἄλλοις·
ἀλλ᾽ εἴτε φῶς εἴτ᾽ ἀήρ ἐστι τὸ μεταξὺ τοῦ ὁρωμένου καὶ
τοῦ ὄμματος, ἡ διὰ τούτου κίνησίς ἐστιν ἡ ποιοῦσα τὸ ὁρᾶν.
καὶ εὐλόγως τὸ ἐντός ἐστιν ὕδατος· διαφανὲς γὰρ τὸ ὕδωρ,
ὁρᾶται δὲ ὥσπερ καὶ ἔξω οὐκ ἄνευ φωτός, οὕτως καὶ ἐντός·
διαφανὲς ἄρα δεῖ εἶναι· ἀνάγκη ἄρα ὕδωρ εἶναι, ἐπειδὴ οὐκ
ἀήρ. οὐ γὰρ ἐπὶ ἐσχάτου τοῦ ὄμματος ἡ ψυχὴ ἢ τῆς ψυχῆς
τὸ αἰσθητικόν ἐστιν, ἀλλὰ δῆλον ὅτι ἐντός· διόπερ ἀνάγκη
διαφανὲς εἶναι καὶ δεκτικὸν φωτὸς τὸ ἐντὸς τοῦ ὄμματος.

In Chapter 3, Aristotle laid out his theory for colour, where he confronted ideas like that of the Pythagoreans, who defined colour as the superficies of an object since it resides at the object's boundaries. Aristotle stated that there is no reason not to suppose that the same natural substance that exists in an object's outer parts (and is the source of the perceived colour), also exists internally and produces the same colour within, so the Pythagorean theory was flawed. Like other thinkers of that time, Aristotle first defined white and black and then all other colours in close relation to substances and ingredients and their combinations and mixtures in a multitude of ratios. There are other references on colour and its perception in Aristotle's *Meteorologica* (or Meteorology), Book III (Aristotle, 1994; Lee, 1952), where he explains the formation of the rainbow and its colourful appearance as a juxtaposition of the main colours, *red, green and purple*[13]:

> ...ἡ μὲν γὰρ ἐντὸς τὴν πρώτην ἔχει περιφέρειαν τὴν μεγίστην φοινικίαν, ἡ δ' ἔξωθεν τὴν ἐλαχίστην μὲν ἐγγύτατα δὲ πρὸς ταύτην, καὶ τὰς ἄλλας ἀνάλογον. ἔστι δὲ τὰ χρώματα ταῦτα ἅπερ μόνα σχεδὸν οὐ δύνανται ποιεῖν οἱ γραφεῖς· ἔνια γὰρ αὐτοὶ κεραννύουσι, τὸ δὲ φοινικοῦν καὶ πράσινον καὶ ἁλουργὸν οὐ γίγνεται κεραννύμενον· ἡ δὲ ἶρις ταῦτ' ἔχει τὰ χρώματα. τὸ δὲ μεταξὺ τοῦ φοινικοῦ καὶ πρασίνου φαίνεται πολλάκις ξανθόν.

---

[13] The passages in Greek that appear in the next paragraphs are adopted from https://goo.gl/kuXEgx.

Chapter 3
Colour Theory During the Middle Ages

3.1 Alhazen

Abū 'Alī al-Hasan ibn al-Hasan ibn al-Haytham[1] (965–1040) or simply Ibn al-Haytham or Al-Hazen or Alhazen (from his Latinised first name, al-Hasan), born in Basra, Buyid Emirate (Southern Iraq), was an Arab mathematician, astronomer and physicist of the *Islamic Golden Age*.[2] Alhazen is considered to be one of the most influential figures in science. He made significant contributions to our knowledge of optics and visual perception and switched the general mindset and misconception of the past regarding the way light and colour are perceived.

In 1021 Alhazen created a lengthy seven-volume treatise, *Kittāb al-Manāzir* (كِتَاب أَلمَنَاظِر), the *Book of Optics* (Alhazen, 1021; Risnero, 1572), in which he contradicted, based on experimentally founded arguments, the widely accepted theory of vision of most of the ancients, and supported a model of vision, in which light enters the eyes rather than being transmitted from them. In this treatise, he discussed experiments regarding the nature of light and how light decomposes into its constituent colours. He also discussed phenomena such as the reflection on mirrors, as well as the refraction of light when passing through different kinds of media. In his theory an object emits rays of light from every point on its surface; these rays travel in all directions, some of which reach the observer's eyes. A significant consequence

---

[1] أَبُو الِى أَلحَسَن إِبن أَلحَسَن إِبن أَلهَيثَم

[2] Typically referring to the period between the 8th and the 13th century.

Fig. 3.1 The structure of the
human eye by Alhazen

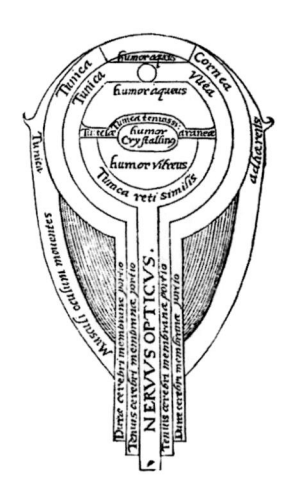

of his line of reasoning is that objects perceived this way can be considered to be
sets of infinite amounts of points.[3]

Alhazen distinguished light into two forms, the primary and secondary, and
explained that the primary or essential light emanates from self-luminous objects
while the secondary or accidental light comes from the objects that capture essential light from self-luminous objects and re-emits it. Thus, accidental or secondary
light may exist only in the existence of a source of primary or essential light. Light
can pass through objects and the amount of light passing through is proportional to
the degree of the opaqueness of the object. Absolute transparency does not exist. In
addition, when light passes through different media it is being refracted or can be
reflected by smooth objects (like mirrors), *always travelling in straight lines*. In his
experiments, Alhazen concluded that colour cannot exist without air.

Alhazen was also particularly concerned with understanding the mechanisms of
visual perception beginning with the formation of the images within the eyes. He
identified an inconsistency in the formation of images in the eyes and his theory of
the infinite rays reflected from the objects; his way out of this inconsistency was that
from those rays only the one that enters the eye perpendicularly is essential for the
sensation of vision, and this is due to the refraction of the rays when entering the eye
that weakens those that enter at an angle other than perpendicular. Figure 3.1 shows
Alhazen's drawing of the structure of the human eye, as illustrated in (Risnero, 1572).

Last but not least, the first clear explanation of a *camera obscura* and the inverted
projection through an aperture is attributed to Alhazen, which is a significant step
in optics and especially in photography. In the *Book of Optics* Alhazen used the
term المتهلم البت (*Al-Bayt al-Muthlim*), which translates to *dark room* to describe
it. What is particularly important about Alhazen is that he is considered to be the
father of modern scientific methodology, due to his emphasis on experimental data
and reproducibility of experimental results.

---

[3] An early *pointillism* from an artistic perspective, or a very early *quantum-ish* view of nature from
a scientific perspective.

3.2 Vitello

Erasmus Ciołek Vitello (or Witelo; c.1270–1285) was a German-Polish natural philosopher from Silesia. He was also a friar and a theologian. He is considered to be an important figure in the Polish history of philosophy and deserves to be included in an account of contributors to the foundations of colour theory due to his *Ten Books of Optics* (originally, *Vitellonis Thuringopoloni opticae libri decem*, in Latin) included in Risner's[4] *Opticae thesaurus* (Risnero, 1572), along with Alhazen's *Book of Optics*. Vitello's work, as presented in (Risnero, 1572) reveals that it is purely based on Alhazen's work (even the figures are the same) and includes Vitello's contribution to Alhazen's work. It has to be noted that Vitello's work had a significant impact on the work of other European scientists, particularly Johannes Kepler, and also influenced the Renaissance theories of perspective. Vitello also discusses matters relating to psychology, ideas about the subconscious, and references Plato and Socrates. He, somehow, proposes that light itself is a sensible entity.

Figure 3.2 shows the cover page from Risner's *Opticae thesaurus* for Vitello's and Alhazen's works. It is a very interesting illustration, in which one may find depictions of the sun as a light source, the mechanisms of reflection and refraction and the formation of the rainbow.

Fig. 3.2 Cover page from Risner's *Opticae thesaurus*

---

[4] Friedrich Risner, or Federico Risnero, or even Fridericus Risnerus (c.1533–1580) was a German mathematician who is known for his 1572 publication of the *Opticae thesaurus* and his proposal of a *portable camera obscura*, as a drawing aid, in the form of a small hut that could be carried around the countryside for landscape painting.

Fig. 3.3 Vitello's stereo
vision diagram from Risner's
Opticae thesaurus

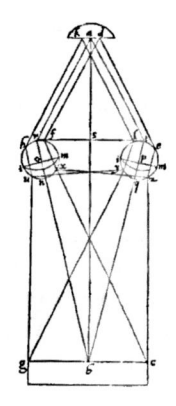

In addition, Fig. 3.3 shows Vitello's diagram for stereo vision and the formation
of the single image from the two images of each of the eyes. As he states, the eyes
collect all points onto their radial surface and drive the image to the nerve.

> Omnes forme punctorum aqualiter circumflantium puncta,
> quae superficiebus uisuum incidunt secundum axes ra-
> diales: ad puncta aqualiter circumflantia medium punc-
> tum nerui communis consimiliter pertingunt.

In (Burchardt, 2004) there is an interesting translation of parts of Vitello's work
regarding the notions of refraction, particularly of colour dispersion through crystal
objects. It should be also noted that Thomas Young in 1807 published his paper *On
the History of Optics*, in which he stated that

> Alhazen was mistaken in some of his propositions re-
> specting refraction; Vitello, a native of Poland, gave a
> more correct theory of this subject, and constructed a
> table of refractive densities, showing the supposed pro-
> portions of the angles of incidence and refraction in the
> respective mediums.

attributing to Vitello the correct view on the issues of the optics discussed by those
two great minds (Young, 1807c).

Chapter 4
The Modern Era of Colour Theory

—Undulations are excited in this ether whenever a body becomes luminous.

Thomas Young, On the theory of light and colours

4.1 Leonardo da Vinci

Leonardo di ser Piero da Vinci, or simply Leonardo da Vinci (1452–1519) was an Italian polymath who is considered among the founders of the Renaissance. His extensive diversity of talents transformed his name into a brand name to characterise those extremely rare individuals in human history to poses an exquisite genius. He is considered both an artist and a thinker, with a pervasive interest in conclusions through observation and experiential cognition. As expected, his interest in light and vision was primarily due to his artistic nature that was also coupled with his fact-seeking impulse through accurate observation.

To Leonardo's mind, the human brain collects visual and other stimuli, processes them into sensory perception and subsequently transmits responses to the muscles through the nerves (Isaacson, 2017). Leonardo's attention to observation details in all his aspects of work seems to have been based on an attempt to highlight and illustrate the superiority of vision among the senses. As Isaacson (2017) quotes Leonardo saying, *the eye, which is said to be the window to the soul, is the primary means by which the sensory receptor of the brain can fully and magnificently contemplate the infinite works of nature*. He was in awe of the function of vision, as evident in his poetic text in the *Optics* of his notebooks (MacCurdy, 1955) (c.a.345 v.b),

G. Pavlidis, *A Brief History of Colour Theory*,
https://doi.org/10.1007/978-3-030-87771-2_4

> Who would believe that so small a space could contain the images of all the universe? O mighty process! What talent can avail to penetrate a nature such as these? What tongue will it be that can unfold so great a wonder? Verily, none! This it is that guides the human discourse to the considering of divine things. Here the figures, here the colours, here all the images of every part of the universe are contracted to a point. O what point is so marvellous! O wonderful, O stupendous Necessity thou by thy law constrainest all effects to issue from their causes in the briefest possible way! These are the miracles, ... forms already lost, mingled together in so small a space, it can recreate and reconstitute by its dilation.

His observations on the formation of shadows are defining aspects of his paintings (as are of his concepts). Due to these observations, his painting practices, mathematical aspects and his theory of optics, he admitted that there are no precise borders of the objects. Through experimentation[1] and anatomy, he maintained that the image of a scene is formed as a whole in the eye, and this is way no clear borders can be imaged.

By having a perfect understanding of the principle of the camera obscura and the inverted image, he was fascinated how the eyes do not suffer from this phenomenon. His way out of this impasse was to propose a double-crossing of the rays in the eye so that eventually the formed image turns upright before moving on to the brain (MacCurdy, 1955; Isaacson, 2017) as shown in the diagram in Fig. 4.1 based on Leonardo's original drawing. According to his theory, the eyes are spherical so that light from multiple angles can be captured and form images of a wide scene (MacCurdy, 1955) (D I r.).

> Nature has made the surface of the pupil situated in the eye convex in form so that the surrounding objects may imprint their images at greater angles than could happen if the eye were flat.

In his notes on the *Optics*, and his notebooks (MacCurdy, 1955), he seems well-aware of the classical laws of light propagation, such as refraction, as revealed in the following text.

> And this process of contraction proceeds from the fact that the rays of the images approach the perpendicular when they pass from the thin to the dense, and that the albugineous humour is

---

[1] For example, based on Alhazen's work, he experimented among others with a needle, which when brought close to the eye starts to become blurry and translucent, and finally disappears just in front of the eye.

Fig. 4.1 Leonardo Da
Vinci's concept of double
ray crossing in the eye

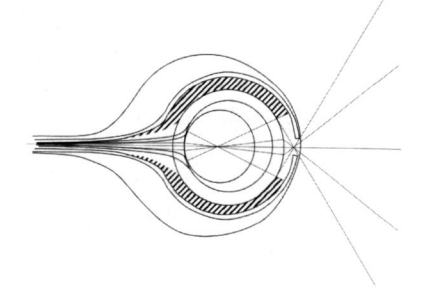

here much thinner and more subtle than the space enclosed by the
surface of the vitreous sphere.

In addition, his observations led him to differentiate central to peripheral vision,

The eye has one central line and all the things that come to the
eye along this line are seen distinctly. Round about this line are
an infinite number of other lines that adhere to this centre line and
these have so much less strength in proportion as they are more
remote from the central line.

In Leonardo's notebooks, there is one about the *Atmosphere*, in which he stated
something quite remarkable for his time. Although to that date, a ray-based, mostly
particle-like theory of light was widely accepted, he somehow imagined a *wave
theory of light* (MacCurdy, 1955; Keele, 1955), in saying that

Just as the stone thrown into the water becomes the centre and
cause of various circles, and the sound made in the air spreads
itself out in circles, so every body placed within the luminous air
spreads itself out in circles and fills the surrounding parts with an
infinite number of images of itself, and appears all in all and all in
each smallest part.

At the same time, he made numerous drawings and anatomical designs (like the one
in Fig. 4.1), in which the ray-based light propagation was also adopted, making it
seem like he already imagined a *dual nature for light*, which was established some
5 centuries later.

Regarding human perception, Leonardo's concepts were clearly expressed in that
there is a hierarchy in human physiology or a process by which the mind is the
decisive component, which is fed with information by the senses (MacCurdy, 1955).

> The soul apparently resides in the seat of the judgment, and the judgment apparently resides in the place where all the senses meet, which is called the common sense; and it is not all of it in the whole body as many have believed, but it is all in this part; for if it were all in the whole, and all in every part, it would not have been necessary for the instruments of the senses to come together in concourse to one particular spot; rather would it have sufficed for the eye to register its function of perception on its surface, and not to transmit the images of the things seen to the sense by way of the optic nerves; because the soul—for the reason already given—would comprehend them upon the surface of the eye. So therefore the articulation of the bones obeys the nerve, and the nerve the muscle, and the muscle the tendon, and the tendon the common sense, and the common sense is the seat of the soul, and the memory is its monitor, and its faculty of receiving impressions serves as its standard of reference.

Nevertheless, he was somehow reluctant to accept the existence of colours (but only of light) and to propose a clear colour theory. He seems to have accepted a complicated view, in which there is only white light and shadows. In his notes about *Painting*, he stated that the medium between the eyes and the objects transforms the view of the objects to appear in a particular colour. Like all translucent or transparent media are to be treated as coloured media which impose the apparent colouring. But then, he also stated that the reason for the formation of a rainbow is the light rays travelling through the water drops in the air; yet in another passage, he definitely concludes that it is not the sunlight that is responsible for the rainbow. He also definitively concludes that the eyes have no share in the creation of colours. In another passage regarding light and colour, he suggested that "the quality of colours becomes known by means of light". In his notes about *Colour* (MacCurdy, 1955) (XXX.Colour), Leonardo focuses on colour differences, particularly colour opponency, apparently through his experience in colour adaptation (particular colour against a completely different background). His ideas about colour influenced Göethe and exhibit a slight resemblance to the opponent colour theory discovered later.

4.2 Johannes Kepler

Johannes Kepler (1571–1630) was a German astronomer and mathematician from the Free Imperial City of Weil der Stadt, Holy Roman Empire (later Germany). Kepler is a central figure in science and particularly for the scientific revolution of the 17th century that he helped set in motion. He is known for his work in astronomy and especially his laws of planetary motion, as laid out in his 1609 book *Astronomia Nova ΑΙΤΙΟΛΟΓΗΤΟΣ seu physica coelestis, tradita commentariis de motibus stellæ Martis ex observationibus G.V. Tychonis Brahe* (or simply *New Astronomy*), where he proposed the elliptical planetary orbits in the heliocentric model and laws regarding the speed and axis of rotation of planetary motion. Apart from his seminal work in astronomy, Kepler was also deeply interested in the study of optics and human vision. He is considered to be the first to accept the inverted-reversed image projection onto the retina of the eye. The correction of the image is done, according to him, in the brain, although he did not seem to be so concerned about how, since his primary interest was in the optics.

Of particular interest is his largely important work published in 1604 under the long title *Ad Vitellionem Paralipomena, quibus Astronomiæ Pars Optica Traditur* (*Paralipomena to Witelo whereby the Optical Part of Astronomy is Treated* or *Paralipomena to Witelo and the Optical Part of Astronomy*) (Kepler, 1604). This treatise is considered to be a foundation of optics in their modern form. He was particularly interested in projective geometry, which he tried to found upon conic sections, while he described how projective space changes would change shapes within this family of conic sections (ellipse becomes a parabola, merging of the two foci of ellipse results in the formation of a circle, merging of the foci of a hyperbola transforms it into a pair of straight lines, a straight line extended to infinity will meet itself at a single point at infinity). A translation of the original Latin text in English can be found in Kepler & Donahue (2000).

Kepler, analysed the nature of light and its propagation, along with its interaction with bodies, but was also concerned about the visual sense mechanism. He expressed his theory on the nature of light in the form of a series of not less than 38 Propositions. In the opening Propositions, Kepler theorises that *light has a specific origin and travels instantly on spherical surfaces that extend radially from the source to infinity, without losing strength; this motion can be envisioned in directions of straight lines, defined by dense rays.* In his own words,

> Luci effluxus vel e iaculatio competit à sua origine in locum distantem...Punctum quodlibet infinitis numero lineis effluit. Scilicet vt orbem omnem circumcirca illustret, quod sieri debere diximus Sphæricum autem infinitas habet lineas...Lux seipsa in infinitum progredi apta est...Lineæ harum eiaculationum rectæ sunt, dicantur radij...Lucis motus non est in tempore, sed in moment.

Kepler explained the instantaneous propagation of light in terms of the Aristotelean law of motion[2] and his hypothesis that *light has no mass,* by which he concluded that time is not involved in the motion of light. Interestingly, in Proposition VIII, he suggests that the light rays should be also be considered as indications of the motion, but what is moving through space is a surface.

---

[2] Kepler expressed it as in terms of time, supporting that time is proportional to the ratio between the moving mass to the medium in which motion takes place, or the ratio of the moving power to the mass.

> *Lucis radius nihil est de luce ipsa egrediente.* Nam radius per IV.
> nihil aliud est nisi ipse motus lucis. Sané vt & in motu physico,
> motus ipsius est recta linea, physicum vero mobile, est corpus: ita
> in luce motus ipse est recta itidem linea, mobile veró, est super-
> ficies quædam. Et vt illic recta motus non pertinet ad corpus, sic
> hic recta motus non pertinet ad superficiem.

In Proposition XV, Kepler mentioned colour for the first time, where he expressed an interesting hypothesis that *colour is light in potentiality, enclosed in a transparent material.* In a way, colour is a quality of matter and the colour on objects includes the potential to transmit light if excited by the light from the Sun. *Differences in light intensity and in material density and transparency are the qualities that define the various colours.*

> Color est lux in potentia, lux sepulta in pellucidi materia: si
> iam extra visionem consideretur; & diversi gradus in dispositione
> materiæ, causa raritatis & densitatis, seu pellucidi & tenebrarum;
> diuersi item gradus luculæ, quæ materiæ est concreta, efficiunt
> discrimina colorum.

Kepler in the subsequent Propositions reaffirmed the laws of reflection and refraction and summarised all the interactions of light with matter in Proposition XXVII. Light in the same medium is partly reflected (Prop. XVIII), partly refracted (Prop. XX), partly adheres to the colour of the medium (Prop. XVI) and attenuated by the colour of the medium (Prop. X).

> Lux in codem medio partim repercutitur, partim infringitur, par-
> tim & in colore medii adhaerescit, seu à colore reuibratur, atque
> ita in tenuiores luces diuiditur.

To Kepler the law of light mixing is additive, and the resulting colour would depend on the proportion of the mixed lights' densities or strengths. Furthermore, he attributed *heat* to be *a property of light*, or a carrier of heat. And as heat in objects is generated in time, longer exposures to light ultimately destroys and burns the objects and bleaches the colours (as colours are properties of the matter). Kepler in the Appendix to his chapter on the nature of light included a detailed critical analysis of Aristotle's theory of light and colour.

Kepler's theory of image formation includes Definitions and Propositions of deep insight, following the tradition of Witelo. To Kepler, an Image is practically nothing in itself, as it is the vision of an object subject to the errors in the sense of vision. An image is similar to an imagination, whereas an object is actually an objective entity. Nevertheless, an image carries important information that includes colour, position, distance and quantity.

> Primum ex Catoptrice, in quam ingredimur, definitionem Imaginis desumptam in vestibulo colloco. Dicunt enim imaginem optici, cùm res ipsa quidem cum suis coloribus & figuræ partibus cernitur, sed situ alieno, alicubi & alienis indura quantitatibus & partium figuræ proportione inepta. Breuiter, imago est visio rei alicuius, cum errore facultatum ad visum concurrentium coniuncta. Imago igitur perse penè nihil est, imaginatio potiùs dicenda. Res est cõposita ex specie coloris vel lucis reali, & quantitatibus intentionalibus...Etenim in imagine sunt hæc quatuor potissimùm, Color, situs seu plaga, distantia, quantitas, subsidiis comprehendantur, explicandum: quamuis eadem Vitellio libro 3. & 4. explicauerit.

Kepler made clear that seeing is receiving, which requires contact ("visio sit passio & passio siat per contactum") between the eye and the image of the light rays. The eye is capable of receiving light and colours because it consists of transparent humours ("ocullus constat humoribus pellucidis: hoc ita que respectu lucis & colorum capax est"). Kepler formalised the stereo vision theory by supporting that *two eyes are needed to provide distance estimates through triangulation*. In addition, the eye has a sense of the viewing angles in which it operates ("ocullum sensum habere angulorum apud se constitutorum").

In a subsequent chapter (*Chapter V. De modo Visionis*), Kepler got involved in defining the function of the human vision, as there was no clear description of the subject at that period. Kepler never performed or attended any practice of dissection and thus he cites the work of eminent physicians and anatomists to base his theory of vision. Regarding the anatomy of the human eye, he cites Felix Platter (1536–1614), Johannes Jessenius a Jessen (1566–1621) and Hieronymus Fabricius ab Aquapendente (1537–1619). Kepler comments on the origins of the word *oculus* from the greek ὀπή → ὄπτεσθαι → ὄψις → ὄμμα, ὀφθαλμός, which conveys the meaning of an opening from the opaque organism to the air

> ...quòd hæ sint rimæ seu aperta foramina, è tenebroso capite in clarum aërem pertingentia;

In addition, he makes clear that there are two eyes in animals not as a means of protective redundancy, but as a distance measuring mechanism. The eyes are in a high location so that more distant objects may be observed since the experienced world is on a sphere. Kepler unfolds a line of reasoning on why the eyes needed to be spherical, or in their location close to the brain, why they are aligned with the horizon, why they are individually protected and why they developed rapid motion capabilities to cover large angles of view. He reminds that it was already known that the brain extends, by means of nerves, to the inner surface of the eyes forming a complex layer ("...ipsa verò cerebri substantia neruum opticum...hæcque ispa

Fig. 4.2 Anatomical
drawings of the human eye
from Kepler's *Paralipomena
to Witelo and the Optical
Part of Astronomy*, attributed
to Felix Platter

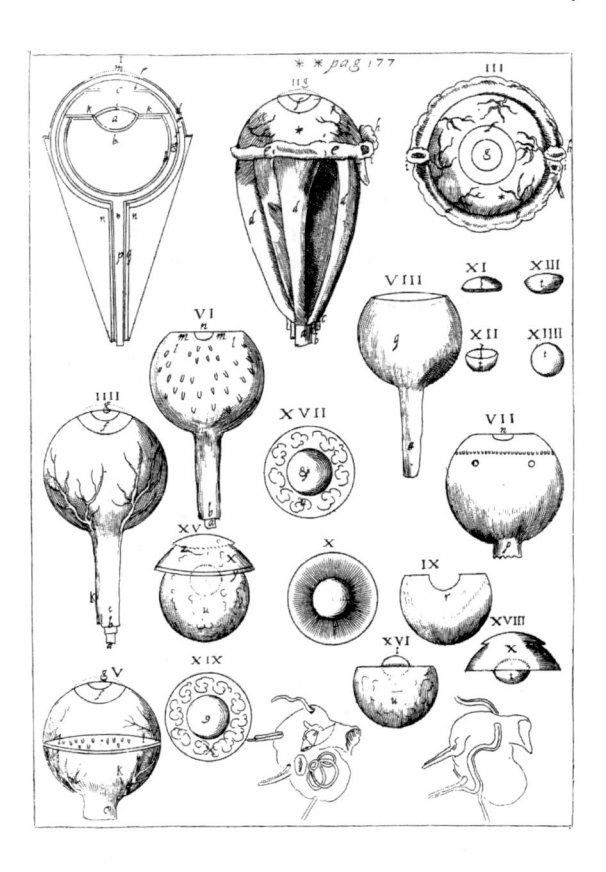

non simplex."). Kepler provided a detailed description of the parts of the eye and
included a figure of multiple anatomical drawings of the human eye, borrowed from
Felix Platter (Fig. 4.2).

Subsequently, in Sect. 2 of Chapter V, Kepler focuses on the means of vision
("modus visionis"), in which he unfolds the complete theory of vision, as he emphatically states, for the first time to his knowledge. Vision occurs when the image of
the whole hemisphere hits the retina of the eye.

> Visionem fieri dico, cum totius hemisphærii mundani, quod est
> ante oculum, & amplius paulò, idolum statuitur ad album sub
> rufum retinæ cauæ superficiei parietem.

He proceeded to confront many of Witelo's propositions regarding the function of
refraction of light in the eyes. Through logical reasoning, he supported that the optical information retrieved by the eyes is altered on its course to the brain, and he
theorised about the usage of the *optic chiasm*, the joining of the left and right eye

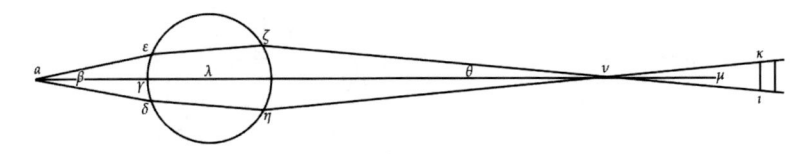

Fig. 4.3 Kepler's diagram that demonstrates the optics of image inversion

nerves halfway to the brain. He specified that *images are inverter in the eyes* and all the colours are being imprinted on the retina.

> Visio igitur sit per picturam rei visibilis ad albam retinæ & cauum parietem; & quæ foris dextra sunt, ad sinistrum parietis lacus sinistra ad dextrum, supera ad inferum, infera ad superum depinguntur: viridia etiã colore viridi, & in vniuersum res quæcunque suo colore intra pingitur...

Kepler envisions the focusing of light by the lens as a simplified graph of two (light) cones with a common base coinciding with the lens of the eye, with the vertex of the one cone on the point of the object being viewed and the vertex of the other on the retina, where the point is being detected. This mechanism works for all points on the objects of the scene in front of the viewer.

Furthermore, Kepler used analogies with crystal balls and vessels filled with clear water to unfold his theory of geometric optics in vision, in the form of 28 Propositions, accompanied with graphs and explanations, particularly to highlight the role of refraction. He also explained the inversion of the images due to rays converging on and diverging from the visual axis in purely geometric terms and the assumption of straight-line ray propagation. Figure 4.3 shows a reconstruction of the original Kepler's diagram that demonstrates the optics of image inversion (Proposition XVII). In the Corollaries of Proposition XVII (and in reference to Fig. 4.3), Kepler detailed how and when images appear inverted, in relation to the location of the observed object or the eyes relative to the points of ray convergence.

> 1. Patet hinc, oculo α longius etiam distante quàm γβ si tamen ιϰ fuerit inter θμ partim euerso situ (in extremis nempe) visum iri, partim (& in intermediis) situ recto: Et candem etiam circularem, iuxta p 43 decimi Vitellionis: itaque confusè.
>
> 2. Quod si sic manente oculo ιϰ sit etiam intra ϑ citimum intersectionis terminum, tota erecta videbitur.
>
> 3. Si verò oculus intra βγ sit, intersectionibus in infinitum excurrentibus, & aliquibus ex oculo radiationibus parallel iter refrac-

tis: Si tunc res in axe fuerit sita, & minor parallelorum distantia, videbitur erecta & euersa simul, siquidem remotior fuerit citima intersectione: Sin propior, erecta tantùm apparebit.

4. At si excesserit complexum parallelorum, ultra terminum tota euersa, medium erectum, & partim circulare apparebit.

5. Deniquibus oculo & re cis terminos intersectionŭ existentib. ille parallelorum, hæc radiationum oculi, res erecta & maximæ quartitatis videbitur.

Kepler made an interesting definition and distinction between an *image* and a *picture*. In the Definition just above Proposition XIX, he assigns the term image to the result of reason, whereas he connects the term picture with the figures of objects on surfaces. Last but not least, he explained the usage of correcting lenses for common eye defects.

Kepler also left a remarkable treatise on dioptrics in his 1611 *Dioptrice*, where one may find his theory arranged in axions, theorems and proposition, along with excellent illustrations (Kepler, 1611).

4.3   René Descartes

René Descartes (1596–1650), born in Touraine France, was a mathematician and philosopher who seemed to share similar philosophical ideas with Aristotle and the Stoics,[3] and, besides his important contributions to mathematics, he is widely recognised as a founding figure of modern philosophy. Apart from the Cartesian geometry, he is best remembered by his quote "cogito, ergo sum", "I think, therefore I am", included in his most famous work *Discourse on the Method* written in 1637 Descartes (1637a, 1667). Descartes supported a *particle theory of light*, by stating that light is made up of discrete *corpuscles* (small particles) that travel in straight lines with a finite speed, thus sparking the modern *corpuscular theory of light*. Apparently, he was influenced by the growing trend towards *atomism* in that period, which was a trend in favour of the ancient atomic theories, as expressed by Leucippus and Democritus.

Descartes was particularly interested in explaining how human vision works as he tried to explain how the human body works in general, in his quest to understand its nature. In his work *Tractacus De Homine* (Descartes, 1677) he draws a particularly

---

[3] The school of the Stoics was a Hellenistic philosophical movement, founded by Zeno of Citium (Ζήνων ὁ Κιτιεύς, c.334–262 BCE) in Athens, during the 3rd century BCE, which supported a philosophy of personal ethics that draws on logic and interpretations of the natural world.

Fig. 4.4 Descartes' graph
of human visual perception
from *Tractacus de homine*
(Descartes, 1677)

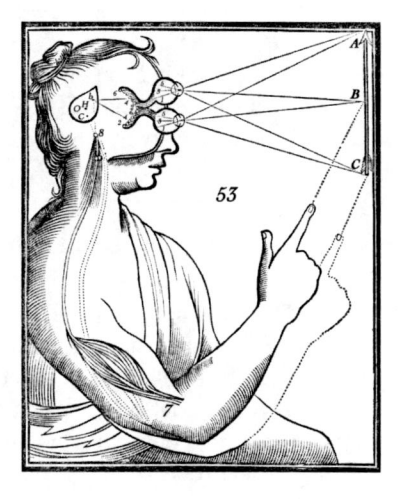

Fig. 4.5 Descartes' graph
of the human eye from *De
homine figuris* (Descartes,
1664)

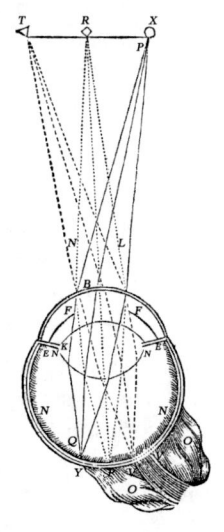

interesting diagram of the human visual perception mechanism, a representation
of which is shown in Fig. 4.4. In addition, in *De homine figuris* (Descartes, 1664),
among others, he displays a graph of the human eye which is shown in Fig. 4.5, on
which Descartes explains the formation of images in the eyes.

In his *Principles of Philosophy* (Descartes, 1644, 1637c) (1644) Descartes
explained that[4]

---

[4] Passage in Latin taken from https://www.loc.gov/resource/rbc0001.2013rosen1431/?sp=1, Pars
quarta. De Terra. CXCV. De visu.

> Denique nervorum opticorum extremitates, tunicam, retinam dic-
> tam, in oculis componentes, non ab aere nec a terrenis ullis cor-
> poribus ibi moventur, sed a solis globulis secundi elementi, unde
> habetur sensus luminis & colorum: ut jam satis in Dioptrica &
> Meteoris explicui.

which can be rendered in English[5] as stating that

> The optic nerves are the organs of the subtlest of all the senses,
> that of sight. The extremities of these nerves, which make up the
> coating inside the eye called the retina, are moved not by air or
> any terrestrial bodies entering the eye but simply by globules of
> the second element which pass through the pores and all the flu-
> ids and transparent membranes of the eye. This is the origin of
> the sensations of light and colours, as I have already explained
> adequately in my Optics and Meteorology.

In *La Dioptrique* (Descartes, 1637b),[6] Descartes begins by stating that

> Toute la conduite de notre vie dépend de nos sens, entre lesquels
> celui de la vue étant le plus universel et le plus noble, il n'y a point
> de doute que les inventions qui servent à augmenter sa puissance
> ne soient des plus utiles qui puissent être.

which translates to that all the conduct of our life depends on our senses, among
which that of sight is the most universal and the noblest; there is no doubt that the
inventions which serve to increase its power are of the most usefulness that can be.
Descartes derives the law of reflection in agreement with what was already known
using a simple example of a ball and the ground. He uses the same logic to derive the
law of refraction by substituting the ground with a thin canvas (a linen sheet), which
the ball can easily penetrate losing only a part of its speed. By assuming that the
speed is only determined by the resistance due to each medium through which the
ball travels, Descartes formulated a proposition that the ratio of the sine of the angle
of incidence to the sine of the angle of refraction is equal to a constant determined
by the resistances of the media in the experiment. This, actually, is *a mechanistically
derived law of refraction*, which raised a lot of controversies and even accusations
of plagiarism (of Snell's law, which was derived around 1621). Overall, Descartes'

---

[5] According to the *Selections from the Principles of Philosophy* found at http://www.
earlymoderntexts.com/authors/descartes, *PART IV. OF THE EARTH. CXCV. Of sight*, found at
http://www.earlymoderntexts.com/assets/pdfs/descartes1644part4.pdf.

[6] A 1657 edition of *Discours De la Methode* that also includes the original texts of *La
Dioptrique* and *Les Meteores* ca be found online @ https://ia600503.us.archive.org/26/items/
discoursdelamet00desc/discoursdelamet00desc.pdf.

view is that rays of light mechanically stimulate the eyes and those mechanical stimulations are then passed to the brain and give rise to perceptual experiences.

4.4 Isaac Newton

Isaac Newton (1642–1726/27) from Woolsthorpe-by-Colsterworth, Lincolnshire, England, was a mathematician, physicist, astronomer and theologian, who was particularly interested in systematically defining light and colour. His 1687 volume titled *PhilosophiæNaturalis Principia Mathematica* (Mathematical Principles of Natural Philosophy) was more or less the foundation of classical mechanics (Newton, 1687).[7]

In 1675 N proposed that light is a continuous flow of particles (the photons) that travel in straight lines, laying foundations for the *corpuscular theory of light* initially suggested by Descartes. The intensity of light is measured by the number of those photons reaching a surface.

Of his first published works on light, colour and visual perception, a letter to the Philosophical Transactions of the Royal Society in 1671 stands out (Newton, 1671), in which Newton laid out his new theory and presented insights from his work on optics and light, in the form of thirteen propositions that connected refraction and colour, the nature of compound colours, the nature of white and the origin of the colour of objects.[8]

As the Rays of light differ in degrees of Refrangibility, so they also differ in their disposition to exhibit this or that particular colour. Colours are not Qualifications of Light, derived from Refractions, or Reflections of natural Bodies (as 'tis generally believed,) but Original and connate properties, which in divers Rays are divers.

To the same degree of Refrangibility ever belongs the same colour, and to the same colour ever belongs the same degree of Refrangibility. The least Refrangible Rays are all disposed to exhibit a Red colour, and contrarily those Rays, which are disposed to exhibit a Red colour, are all the least refrangible: So the most refrangible Rays are all disposed to exhibit a deep Violet Colour, and contrarily those which are apt to exhibit such a violet colour, are all the most Refrangible. And so to all the intermediate colours in a continued series belong intermediate degrees of refrangibility. And this Analogy 'twixt colours, and refrangibility, is very precise and strict; the Rays always either exactly agreeing in both, or proportionally disagreeing in both.

---

[7] One may find copies of the *Principia* in English online. A historical 1729 edition can be accessed online @ https://archive.org/details/bub_gb_Tm0FAAAAQAAJ/mode/2up.

[8] The quoted text from Newton's *New Theory about Light and Colours* is from the Newton Project page @ http://www.newtonproject.ox.ac.uk/view/texts/normalized/NATP00006.

Fig. 4.6 Reproduction of
Newton's figure on the
refraction through a prism
(AX. VIII. in the *Opticks*
states that the object *D* seen
through a prism appears to
be in position *d* due to
refraction)

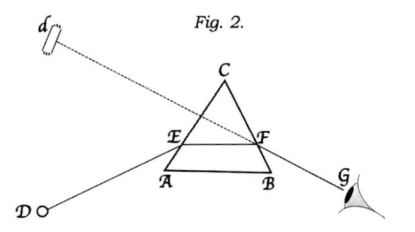

There are therefore two sorts of Colours. The one original and
simple, the other compounded of these. The Original or primary colours are, Red, Yellow, Green, Blew, and a Violet-purple,
together with Orange, Indigo, and an indefinite variety of Intermediate gradations.

But the most surprising, and wonderful composition was that of
Whiteness. There is no one sort of Rays which alone can exhibit
this. 'Tis ever compounded, and to its composition are requi-
site all the aforesaid primary Colours, mixed in a due proportion...Whiteness is the usual colour of Light; for, Light is a confused aggregate of Rays indued with all sorts of Colours, as they
are promiscuously darted from the various parts of luminous bodies. And of such a confused aggregate, as I said, is generated
Whiteness, if there be a due proportion of the Ingredients; but
if any one predominates, the Light must incline to that colour; as
it happens in the Blew flame of Brimstone; the yellow flame of a
Candle; and the various colours of the Fixed stars.

Newton presented a more concrete version of his light and colour theory in
Opticks (Newton, 1704), where he connected refraction with colour through experiments with prisms (Fig. 4.6). In his view, the degree of refrangibility corresponds
with a one-to-one relation to colour, creating a necessary and sufficient condition.
Thus, there is only one colour that corresponds to the same degree of refrangibility.[9]

DEFIN. VII. The Light whose Rays are all alike Refrangible,
I call Simple, Homogeneal and Similar; and that whose Rays
are some more Refrangible than others, I call Compound, Het-
erogeneal and Dissimilar. The former Light I call Homogeneal,
not because I would affirm it so in all respects; but because the
Rays which agree in Refrangibility, agree at least in all their other
Properties.

---

[9] The quoted text from Newton's *Opticks* is from the Newton Project page @ http://www.
newtonproject.ox.ac.uk/view/texts/normalized/NATP00033.

Fig. 4.7 Reproduction of Newton's drawing for his third experiment on support of Proposition II, Theorem II (that the sunlight consists of differently refrangible rays)

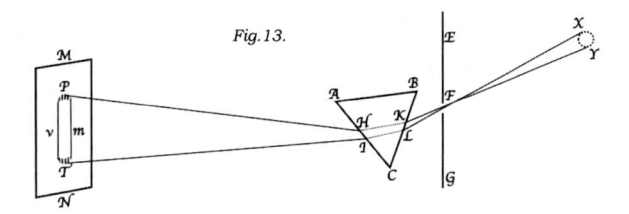

Newton defined all light rays that are alike refrangible as *homogeneal*, whereas light rays differently refrangible as *compound*. In addition, he defined the *primary* colours, like those which correspond to homogeneal light.

> **DEFIN. VIII.** The Colours of Homogeneal Lights, I call Primary, Homogeneal and Simple; and those of Heterogeneal Lights, Heterogeneal and Compound. For these are always compounded of the colours of Homogeneal Lights.

Newton further experimented on the relation between colour and refraction and presented conclusive experiments to support various theorems and propositions.

> **PROP. I. Theor. I.** Lights which differ in Colour, differ also in Degrees of Refrangibility.

Based on those experiments he also performed a number of experiments to prove that the light of the sun consists of rays differently refrangible, thus being compound. Referring to PROP. II. Theor. II., which states that (*the light of the sun consists of rays differently refrangible*), Newton in his third experiment, (graphically depicted in Fig. 4.7) defined the colours in terms of the refrangibility.

> This Image or Spectrum PT was coloured, being red at its least refracted end T, and violet at its most refracted end P, and yellow green and blue in the intermediate spaces. Which agrees with the first Proposition, that Lights which differ in Colour do also differ in Refrangibility.

In addition, in PROP. VI, PROB. II, which tackled the issue of the identification of a compound colour based on a given mixture of known primary colours,[10] presented his colour circle and how its geometrical structure can be used to find out a compound colour (Fig. 4.8). In the example illustrated in this figure, the small circles labelled with small letters p, q, r, s, t, u, v, x represent the gravity centres of

---

[10] In his own words: "In a mixture of primary Colours, the quantity and quality of each being given, to know the Colour of the compound.".

Fig. 4.8 Reproduction of
Newton's colour circle;
drawing adapted from the
Opticks

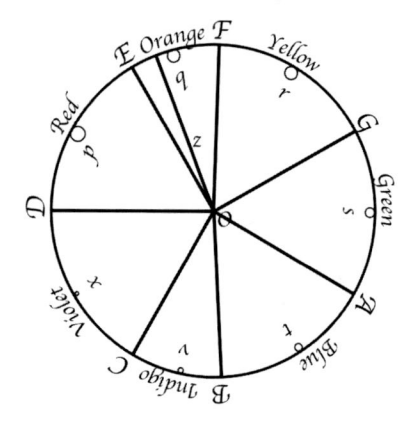

the respective arcs that are defined by the capital letters and correspond to the various colour regions (i.e. the arc DE corresponds to the region of red colours). In addition, the size of the small circles corresponds to the proportion of each colour to the mixture in the example that is being considered. The location of the centre of gravity of these circles is found in the location marked as z. Apparently, z is in the region of the orange colours and a bit to the side of reds (rather than the side of yellows). In addition, as z falls closer to the centre of the circle, which corresponds to the region of least saturated colours, the mixing is thus inferred to correspond to a *relatively faint, slightly reddish-orange compound colour*.

4.5   Christiaan Huygens

Christiaan Huygens (1629–1695) was a Dutch physicist, mathematician, astronomer and inventor. Huygens is appreciated as one of the greatest scientists of all time. He was involved in mechanics, optics, and astronomy, with his astronomical observations of the rings of Saturn and the discovery of Saturn's moon Titan being among the most important contributions. He was among the founders of mathematical physics and a pioneer in the theory of optics. Huygens also improved the design of the telescope by introducing what was subsequently known as the Huygenian eyepiece. One of his most pervasive inventions was the pendulum clock. He left a significant work on mechanics, including the geometrical derivation of the centripetal force and the laws of elastic collision.

What is relevant to this treatise is his significant work in optics. Although he was earlier to Newton, his theory of light and optics was published after Newton's theory. In his *Traité de la lumière* (*Treatise on Light*), published in 1690,[11] Huygens described in detail a new *wave theory of light* (Huygens, 1690, 1912; Huygens et al., 1900). In this treatise, Huygens begins by adopting *a finite speed for light*, as suggested by *Ole Christensen Rømer* (1644–1710, Danish astronomer) after he executed several astronomical observations in the 1670 s. In particular, Rømer observed

---

[11] The treatise begins by stating that the theory was presented to l' Academie Royale des Sciences (the Royal Academy of Sciences) already in 1678.

about 140 eclipses of Jupiter's moon Io for several months while in Uraniborg near Copenhagen, whereas *Giovanni Domenico Cassini* (1625–1712, Italian mathematician, astronomer and engineer) observed the same eclipses in Paris. Rømer worked with Cassini to further investigate the phenomena, which led to a hypothesis for a finite speed of light. Rømer continued to pursue this hypothesis after joining *Jean Picard* (1620–1682, French astronomer) in a collaboration that resulted in a presentation in the French Academy of Sciences in 1676. As in any other case, many scientific breakthroughs had to coincide for a new theory to emerge. It was in 1671 that Jean Picard published his treatise *Mesure de la terre*, where he laid out the foundations of geodesy by using triangulation to measure the diameter of the Earth (Picard, 1671). This led others to describe astronomical distances in terms of Earth diameters. This is most evident in Rømer's and also Huygens work. Through Rømer's estimates, an expression of the ratio of the speed of light to the speed of one Earth orbit around the Sun was $(365 \cdot 24 \cdot 60)/(\pi \cdot 22) \approx 7605$.[12]

Huygens' *Traité de la lumière* consists of six chapters, including the general theory of propagation in a homogeneous medium, the phenomena in media interfaces like reflection, refraction, atmospheric refraction, or special cases like birefringence ("strange refraction of Iceland crystals") and transparent bodies. The basis of the wave theory of light proposed by Huygens is that light propagates as a wave, which corresponds to the activations and collisions of the particles of the ether (the all-pervading medium). Luminous objects are responsible for the propagation of multiple waves as shown in Huygens depiction of a candle flame in Fig. 4.9, in which three distinct points are shown to create three distinct lightwaves. Every point of the candle flame is supposed to create its own wave. And what happens subsequently is that every particle of the ether that is excited becomes a source of further excitations, which are all spherical waves. In a way, every excited point becomes a light source and this is the way light propagates, as shown in Fig. 4.10. Overall, the progressing wavefront still exhibits a spherical nature, not due to the initial excitation but due to the interference between the numerous waves by each of the excited etherial particles, thus the apparent propagation in straight lines is also explained.

> Il y a encore à considerer dans l' emanation de ces ondes, que chaque parcicude de la matiere, dans laquelle une onde s'etend, ne doit pas communiquer son mouvement seule-ment à la particule prochaine, qui est dans la ligne droite tireé du point lumineux; mais qu'elle en donne aussi necess-airement à toutes les autres qui la touchent, & qui s'opposent à son mouvement.

After laying out the general theory of light propagation in a homogeneous medium, Huygens considered the cases of media interfaces, or what happens when light traverses different media. He analysed how his wave theory of light explains the phenomenon of reflection by following the reasoning of a sequential generation of spherical waves. In this case, the geometry of the setup for a reflection preserves equality of angles (of incidence and reflection). The case of reflection is shown in

---

[12] Compare this to the modern estimate of $299792458ms^{-1}/29780ms^{-1} \approx 10066.91$.

Fig. 4.9 Huygens'
depiction of the light waves
from a candle flame

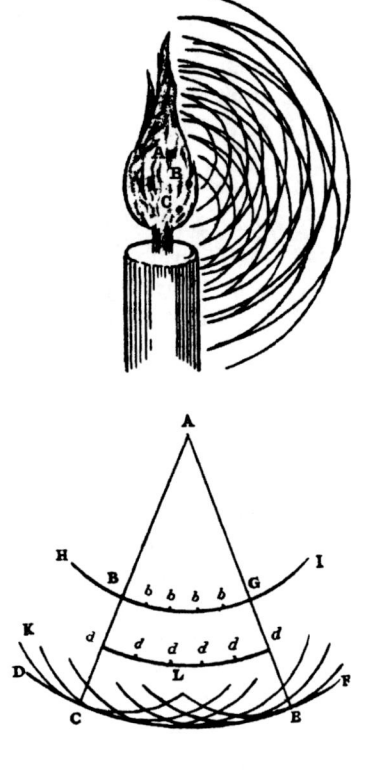

Fig. 4.10 Huygens'
depiction of the propagation
of light waves

Huygens' diagram in Fig. 4.11; as parallel rays (from infinity) hit the reflective surface AB they form spherical waves that at the time the ray at point C reaches point
B on the surface, all waves will have line BN as their a common tangent, thus BN
represents the propagation of the wave AC at that particular moment. Of course, tri-
angles ACB and BNA are necessarily equal—they are both rectangular, have AB as
common side and CB = NA—thus the angles CBA and NAB are also equal. Since
CB marks the direction of incidence and AN the direction of reflection, this means
that the angle of incidence is equal to the angle of reflection. In a similar way, Huygens proceeds to explain refraction, as depicted in Fig. 4.12. For this case, he first
makes clear, that in presence of a medium with a dense substance the particles of the
medium act exactly as the particles of the ether. The slowing of light in the denser
medium results in the phenomenon of the bending of the light rays following the
same reasoning used for reflection. The tangent BN represents the wavefront in the
dense medium just like AC represented the wavefront in the initial rarer medium.
The direction AN, perpendicular to BN is the direction of the refraction, satisfying
the law of refraction, by which the new change in direction depends on the ratio of
the density (refractive indices) of the media.

Huygens' wave theory of light provided elegant explanations to light phenomena,
quite differently than how the particle (or corpuscular) theory of light by Newton

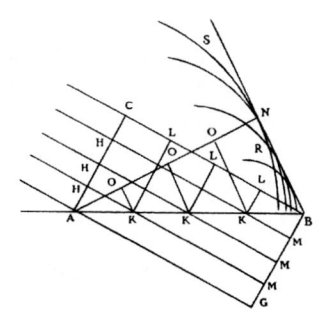

Fig. 4.11 Huygens'
depiction of reflection

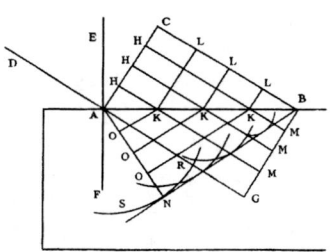

Fig. 4.12 Huygens'
depiction of refraction

could. In some cases, like in refraction, Newton's theory supposed forces acting on
the particles of light to make it bend, but this would result if the speed of light in
denser media were to be increased, which is the exact opposite to what the wave
theory supports. In addition, there is no explanation for diffraction by the particle
theory, whereas a wave theory readily explains the phenomenon. Of course, at the
time those theories were proposed, there was no technology to disprove one or the
other.

4.6 Immanuel Kant

Immanuel Kant (1724–1804), from Königsberg, Eastern Prussia, (Kaliningrad, Rus-
sia) was a German philosopher and one of the founders of the *Enlightenment*. His
most important work is the *Critique of Pure Reason* Kant (1787), in which, in
section *Of Space* (in Section I-of Transcendental Aesthetics states), Kant states that
(from the English translation of J. M. D. Meiklejohn, 1900)

> Colours are not qualities of a body, though inherent in its intuition,
> but they are likewise modifications only of the sense of sight, as it
> is affected in different ways by light.

On the core body-world problem (the basic question of the separation of the inner
and the outer world), Kant argues that space is a priori, therefore man's familiarity
with space is also a priori. However, this is something that he is not accepting for

the objects in space, the experience of which is a property of our aesthetics. In his transcendental idealism, he argued that there are two worlds, that of experiences (thoughts, feelings, and sensory experiences of material things) and that of things not experienced through some known sense, while in this separation the body plays a dual role, as it is a material 'thing' of the world and at the same time is part of the self and a means of perceiving other things.

4.7   George Palmer

During the late1770 s and1780 s, *George Palmer* (1746–1826, or 1740–1795), among various short publications, issued two books on light, colour and colour vision with a long-lasting contribution to the development of colour science. Palmer based his research on physics and chemistry in an attempt to resolve the differences between the colour of light and the colour of objects. In his *Theory of Colours and Vision* (Palmer, 1777), he introduced a theory of colour and vision in a rather interesting form of a dialogue. He also included new experiments with prisms and new explanations of the results. This treatise opens with a list of his *seven principles of colour*[13]

---

1. La lumière ne comporte aucune couleur.
2. Chaque rayon de lumière est compose seulement de trois autres: dont un est analogue au jaune, l'autre au rouge, & l'autre au bleu.
3. Ces rayons sont dans des proportions differentes; & les conservent exactemet, malgre l'accroissement, ou l'affoiblissement de leur rayon principal.
4. Les corps colorées absobent les rayons analogues aux couleurs qu' ils nous presentent, & ne sont appercus que par les autres rayons qu'ils reflechissent.
5. Une surface blanche, réfléchissante toute la lumiere, offre une negation absolue de couleurs.
6. Une surface composee de trois principes colorans, dans une proportion & une intensite convenables, absorbant ces trois rayons conformement au quatrieme principe, offre une negation absolue de lumiere, & un noir parfait.
7. Un feut de ces trois principes colorans peut separement approcher du noir, sans changer de nature, & absorder ses rayons qui ne lui sont pas analogues, losque son intensite encede la proportion de son propre rayon.

---

[13] The quoted principles are from the french version of this treatise that is available as a digitised book in Google Play Books Store @ https://bit.ly/3wY7zuU.

The principles of colour translated in English from the French text, with an influence by MacAdam (1970), can be rendered as follows:

1. Light has no colour.
2. Each ray of light is composed of only three others, one of which is analogous to yellow, the other to red, and the other to blue.
3. These rays are in different proportions; which are kept exactly, despite the increase, or the decrease of their main ray [luminance].
4. The coloured objects absorb the rays analogous to the colours which they present to us, and are only perceived by the other rays which they reflect.
5. A white surface, reflecting all the light, offers an absolute negation of colours.
6. A surface painted with the three primary colourants, in a suitable density, by absorbing the three rays in accordance with the fourth principle, offer an absolute negation of light, and a perfect black.
7. One of these three colouring principles can separately approach black without changing its nature, by absorbing rays not analogous to it, when its density exceeds the proportion of its own ray.

Palmer's theory suggests a basis of three primary colours, which in this case are *red, yellow and blue*. As in similar theories, all possible rays of light contain different proportions of the primary colours. The colours of objects appear so due to the absorption of the rays relating to their colour and the reflection of the other rays. According to the seven principles, rejection of all light indicates a white surface. On the other side, black is created when the three colouring primaries absorb the rays of other colours, creating an intensity that exceeds the proportion of the colour. Most importantly, *light has no colour*; it is that coloured surfaces absorb rays and white surfaces reflect them.

In *Théorie de la lumière applicable aux arts et principalement à la peinture* (Palmer, 1786), Palmer extended his reasoning presented in *Theory of Colours and Vision* with a theory of the prism, in an attempt to clarify and support his idea of the *destructive nature of light*, while enhancing, even more, the role of chemistry in the creation of colour. Palmer clarified that his attempt was towards a unified description for colour, enough to explain colour and light for the arts and science (chemistry and physics). One of his major concerns was to improve the arts by connecting them to science.

Palmer's most interesting contribution was the speculation that there are three different mechanisms in the human eye that account for colour vision. As he states in the *Theory of Colours and Vision*

> The superficies of the retina is compounded of particles of three different kinds, analogous to the three rays of light; and each of these particles is moved by its own ray.

This is a preliminary statement of a trichromatic colour vision, posed about twenty-five years before any other such statement was eventually made by Thomas Young.

4.8 Goethe

Johann Wolfgang von Goethe (1749–1832) was born in the Free Imperial City of Frankfurt, Holy Roman Empire (Germany). Among many things, Goethe (or Göthe) was a poet, playwright, and scientist. He is considered to be the greatest German literary figure of the modern era. Apart from the novels, poems, dramas and other types of literature he left and is most known for, Goethe was also concerned with natural science and particularly of the nature of colours, for which he wrote a treatise. He published in 1810 his most important work titled *Zur Farbenlehre* (von Goethe, 1810), which was in 1840 translated to English and published as *Theory of Colours* (Goethe, 1840), which can be found summarised in (Zajonc, 1976).

Goethe was an opponent of Newton's theory of colour, and was particularly interested in an aesthetic approach towards the understanding of the colour phenomena, with little, though, relation to analytic and mathematical analysis. Most important are his observations on the effect of opponent colours, which led him to redefine and arrange the colours of the typical colour wheel so that three pairs are in diametrically opposite positions, yellow opposite to violet, orange opposite to blue, green opposite to red (as described and depicted in Part I —Physiological Colours, Section V. Coloured Objects, par. 50 of his Theory of Colours). His main concern was to establish a theory that would be consistent with the *qualitative* evaluation of the perception of colour phenomena, one in which perception and explanation would be at the centre. What triggered his divergence from Newton's approach and the pursuit of an alternative theory was his (supposed) discovery that Newton's prismatic experiment was erroneous in that there is no green colour directly exiting a prism, but green is rather a composition of yellow and blue only after some distance from the prism, where the two colours overlap, as shown in Fig. 4.13.[14] Apparently, this observation is false and depends on the topology of the experiment and typical light effects at boundaries.

Ultimately, Goethe considered *darkness* not the absence of light, but rather, a second elusive entity that interacts with light to produce colours. In his experiments with black-white edges he pointed out that yellow appears in the white region of the edge, whereas blue appears in the dark region of the edge, and their mixture produces the green. The darkening of the two basic colours yellow and blue gives rise to reddish hues. Overall he proposed two alternative colour models, either a three-colour model based on yellow, red and blue or a six-colour model based on yellow, red, blue, green, white and black.

---

[14] The reproduction was created based on the original drawing in Plate IV., Fig. 1, of the 1840 English edition found @ https://archive.org/details/Goethe_theory_of_colours_prism.

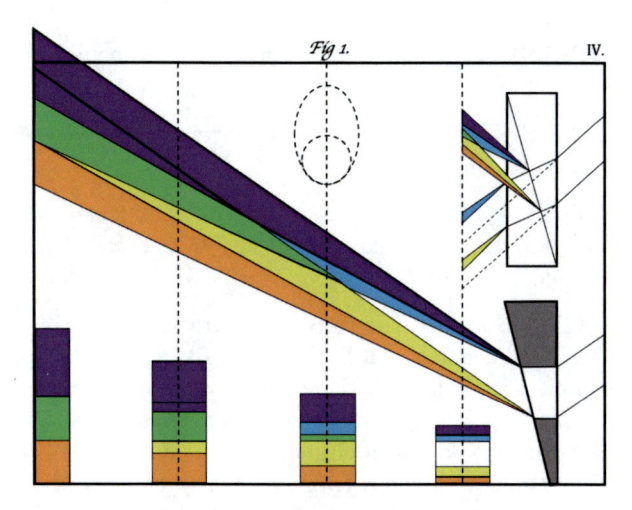

Fig. 4.13 Reproduction of Goethe's fundamental observation on the light dispersion in prisms

Essentially, Goethe rejected the 'sterilised' approach of a theory of colour, in which colour is deprived of its sensation, and is treated as an objective phenomenon even without the need for an observer to experience it. He was deeply certain that talking about colour has no meaning outside of the context of its perception, through the active sensation of vision.

4.9   Thomas Young

The period followed Newton's theories, *Thomas Young* (1773–1829) from Milverton, Somerset, England, was among the first great minds to have also tried to define light, colour and vision. His major contribution in the domain is considered to be in two treatises, *On the theory of light and colours* Young (1802) and *A course of lectures on natural philosophy and the mechanical arts* (Young, 1807a, b), some of his contributions also found in Huygens et al. (1900). In these treatises, he presented his main conclusions regarding colours, following the usual line of reasoning that accepts some of them being primary and the other being combinations of the primaries. Young was among those scientists to support a *wave theory of light*, in favour of Christian Huygens' view and against Newton's particle theory, which was the pervasive theory of that time. Young, although influenced by the Newtonian theory, formulated a theory, which he believed to be opposite to that of Newton and rather close to Christian Huygens' view. His theory was based on a set of hypotheses, largely influenced by Newton, beginning with the existence of the *luminiferous ether* that is supposed to be a rare and elastic medium pervading the universe. These four fundamental hypotheses are,

HYPOTHESIS I. A luminiferous ether pervades the uni-verse, rare and elastic in a high degree.

HYPOTHESIS II. Undulations are excited in this ether whenever a body becomes luminous.

HYPOTHESIS III. The sensation of different colours depends on the different frequency of vibrations excited by light in the retina.

HYPOTHESIS IV. All material Bodies are to be considered, with respect to the Phenomena of Light, as consisting of Particles so remote from each other, as to allow the ethereal Medium to pervade them with perfect freedom, and either to retain it in a stale of greater density and of equal elasticity, or to constitute, together with the Medium, an Aggregate, which may be considered as denser, but not more elastic.

Young preferred the term *undulations* instead of *vibrations* because he wanted to differentiate from the classical view of what vibration is—a permanent forward and backward motion. Instead, an undulation is something that traverses space. This is apparent when he talks about vibrations in the retina of the eye in Hypothesis III. In his scholium (comment) on this Hypothesis he laid out a theory by which there is a limited number of particles (substances) in the retina capable of capturing the undulations of light; he referenced, although as an example, the existence of three such particles, capable of capturing the three principal colours, which he named to be red, yellow, blue. The sensation of all colours comes as the appropriate combination of matching light undulations to retina vibrations. Thus Young's 1802 Bakerian Lecture Young (1802) references *three primary colours, red, yellow and blue* and *transforms the problem of colour perception from an objective natural phenomenon to a subjective human function, by which three receptors are responsible for the tridimensionality of colour vision.*

Young called Hypotheses I–III as essential hypotheses and only provided comments and thoughts on them. It is Hypotheses IV that is totally unique to Young, which differentiated his view from that of Newton's and aligned with that of Huygens. He presented this latter hypothesis extensively with several Propositions, Scholia and Corollaries. In summary, his theory stated that

PROPOSITION I. All are in an elastic Medium Impulses propagated homogeneous with an equable Velocity.

PROPOSITION II. An Undulation conceived to originate from the Vibration of a Particle, must expand through a homogeneous

Medium single in a spherical Form, but with different quantities of Motion in different Parts.

PROPOSITION III. A Portion of a spherical Undulation, admitted through an Aperture into a quiescent Medium, will proceed to be further propagated rectilinearly in concentric Superficies, terminated laterally by weak and irregular Portions of newly diverging Undulations.

PROPOSITION IV. When an Undulation arrives at a Surface which is the Limit of Mediums of different Densities, a partial Reflection takes place, proportionate in Force to the Difference of the Densities.

PROPOSITION V. When an Undulation is transmitted through a Surface terminating different Mediums, it pro-ceeds in such a Direction, that the Sines of the Angles of Incidence and Refraction are in the constant Ratio of the Velocity of Propagation in the two Mediums.

PROPOSITION VI. When an Undulation falls on the Surface of a rarer Medium, so obliquely that it cannot be regularly refracted, it is totally reflected, at an Angle equal to that of its Incidence.

PROPOSITION VII. If equidistant Undulations be supposed to pass through a Medium, of which the Parts are susceptible to permanent Vibrations somewhat slower than the Undu-lations, their Velocity will be somewhat lessened by this vibratory Tendency; and, in the same Medium, the more, as the Undulations are more frequent.

PROPOSITION VIII. When two Undulations, from different Origins, coincide either perfectly or very nearly in Direction, their joint effect is a Combination of the Motions belonging to each.

PROPOSITION IX. Radiant Light consists in Undulations of the luminiferous Ether.

Proposition II makes clear enough that light is to be considered as a spherically expanding wave in space. Furthermore, Proposition III describes how apertures influence the propagation of light, which he presented both verbally and graphically,

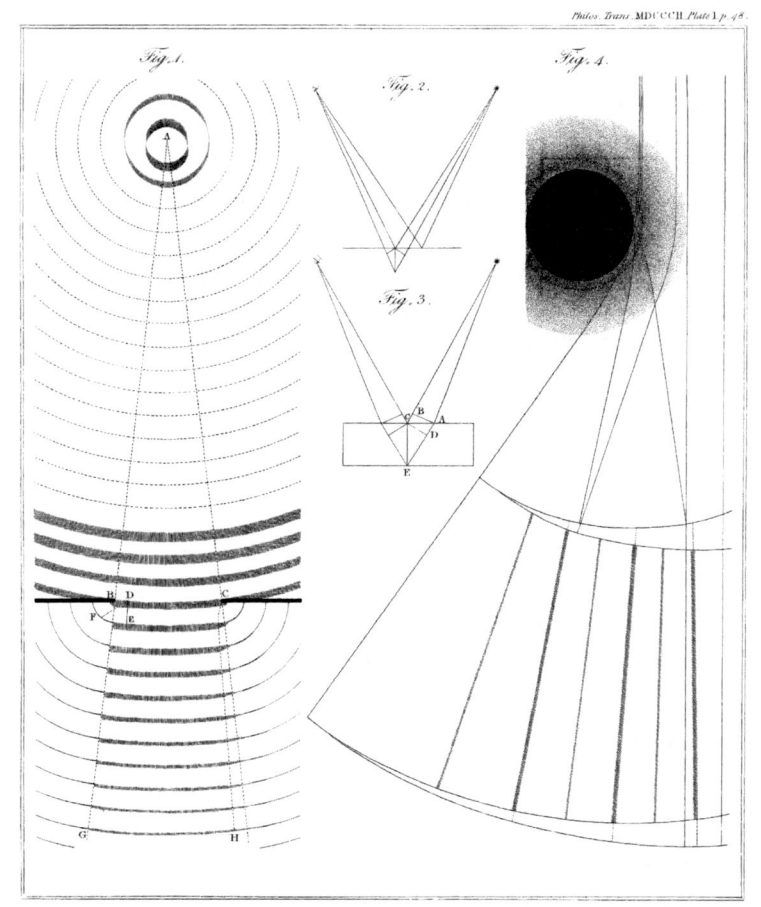

Philos. Trans. MDCCCII. Plate I. p. 48.

Fig. 1. The progress of a series of undulations admitted through an aperture.
Fig. 2. The difference of the paths of the light reflected from two points situated near each other.
Fig. 3. The difference of the paths of the light reflected from the opposite surfaces of a thin plate.
Fig. 4. The paths of two portions of light supposed to pass through an inflecting atmoshpere.

Fig. 4.14   Thomas Young's illustration of light propagation

using Plate I/Fig. 1 of the treatise (shown in Fig. 4.14).[15] Similarly, Propositions IV and V, although not analysed, they are graphically depicted in Plate I/Figs. 2 and 3 (Fig. 4.14), in which Fig. 3 shows refraction in a medium rarer than the surrounding mediums. Proposition VI is an expression of the law of internal reflection. Proposition VII is an expression of the phenomenon of the dispersion of colours, by which light of short wavelength is more refrangible than long-wavelength light. Furthermore, Proposition VIII is a statement of the appearance of interference, which Young analyses in detail using examples for striated surfaces (like gratings), thin

---

[15] This figure was taken from *The Bakerian Lecture: On the Theory of Light and Colours* that is available online @ https://archive.org/details/jstor-107113/page/n1/mode/2up.

Table 4.1 Young's table of colours and frequencies, augmented with modern equivalent representations

Colours	Length of an Undulation in parts of an Inch, in Air.	Number of Undulations in an Inch.	Number of Undulations in a Second. Mil. of Millions.	Wavelength in nm	Wavenumber in cm^{-1}
Extreme	0.0000266	37640	463	675.64	14800.78
Red	0.0000256	39180	482	650.24	15378.94
Intermediate	0.0000246	40720	501	624.84	16004.10
Orange	0.0000240	41610	512	609.60	16404.20
Intermediate	0.0000235	42510	523	596.90	16753.22
Yellow	0.0000227	44000	542	576.58	17343.65
Intermediate	0.0000219	45600	561	556.26	17977.20
Green	0.0000211	47460	584	535.94	18658.81
Intermediate	0.0000203	49320	607	515.62	19394.13
Blue	0.0000196	51110	629	497.84	20086.77
Intermediate	0.0000189	52910	652	480.06	20830.73
Indigo	0.0000185	54070	665	469.90	21281.12
Intermediate	0.0000181	55240	680	459.74	21751.42
Violet	0.0000174	57490	707	441.96	22626.48
Extreme	0.0000167	59750	735	424.18	23574.90

plates of rarer media, as well as thick plates, and provides more insight on the notion of blackness and the colours by inflection, of which he provided Plate I / Fig. 4 (Fig. 4.14) to demonstrate how a varying density medium, like an atmosphere, could inflect the path of light. Young in his discussion on thin plates provides a table of colours and corresponding wavelengths, wavenumber and frequencies, using the estimate for the speed of light of that period, which he states to be expressed as a distance of $500, 000, 000, 000$ feet traversed in $8\frac{1}{8}$ minutes. This is an equivalent $312, 615, 384.62$ m/s.[16] Table 4.1 shows Young's table of colours, augmented with two additional columns that represent the equivalent 'modern' quantities used to describe light and colour (wavelength and wavenumber). His last Proposition IX is a general conclusion of an underlying wave theory of light, which he further supported with additional evidence.

Thomas Young, in his *A course of lectures on natural philosophy and the mechanical arts* (Young, 1807a, b) changed his set of *primary colours to red, green and violet* (now referenced as 'primitive colours'), as he soon observed that his previous colour basis was not of independent colours. In Volume I (Young, 1807a), in *Lecture XXXVII. On Physical Optics*, Young stated that

> It is certain that the perfect sensations of yellow and of blue are produced respectively, by mixtures of red and green, and of green and violet light, and there is reason to suspect that those sensations are always compounded of the separate sensations combined: at least this supposition simplifies the theory of colours;

---

[16] Compare this to the recent estimate of $299, 792, 458$ m/s.

it may, therefore, be adopted with advantage, until it be found
inconsistent with any of the phenomena; and we may consider
white light as composed of a mixture of red, green, and violet,
only, in the proportion of about two parts red, four green, and one
violet, with respect to the quantity or intensity of the sensations
produced.

If we mix together, in proper proportions, any substances exhibit-
ing these colours in their greatest purity, and place the mixture
in a light sufficiently strong, we obtain the appearance of perfect
whiteness; but in a fainter light the mixture is grey, or of that hue
which arises from a combination of white and black; black bodies
being such as reflect white light but in a very scanty proportion.

In addition, in Volume II (Young, 1807b), in *Section VII. Of Dioptrics and Catoptrics,
403. Definition*, Young stated that

Light is distinguished by its effect on the sense of vision, into
white and coloured light; and coloured light into a great number
of various hues: but they may all be referred to the three primitive
colours, red, green, and violet.

4.10   Augustin-Jean Fresnel

Augustin-Jean Fresnel (1788–1827) was born in Normandy, France. He was a civil
engineer and physicist. His research in optics had a significant impact on pushing
towards a wave theory of light for almost 100 years, against Newton's corpuscular
theory. He is most known for his work with lenses, and particularly the invention of
the catadioptric (combination of reflective and refractive) lens. His work on lenses
led to a significant improvement of the visibility of lighthouses, with an important
impact on safer maritime travels. A simpler version of that composite lens, the dioptric stepped lens (refractive), has been widely used in overhead projection devices
and screen magnifiers. The ingenious invention of the catadioptric lenses revolu-
tionised the technology of lighthouses by limiting the propagation of light into a
region that is most important for maritime applications. Typically, light propagates
spherically, which means much of the light travels towards the ground and the sky;
this part is of no practical use. Fresnel designed his composite lens system so that it
exploits refraction and reflection in a way that all light rays are forced to propagate
horizontally (parallel to the ground). Figure 4.15 shows a graph of the principle of
operation of the composite Fresnel lens system, adapted from the 1881 E. Atkinson's English translation of Adolf Ganot's *Cours Élémentaire de Physique* (*Natural
Philosophy*) (Ganot, 1881). Based on the same edition, Fig. 4.16 shows a sketch of
a lighthouse employing Fresnel's catadioptric system.

Fig. 4.15 Fresnel's
catadioptric system for
parallel light rays

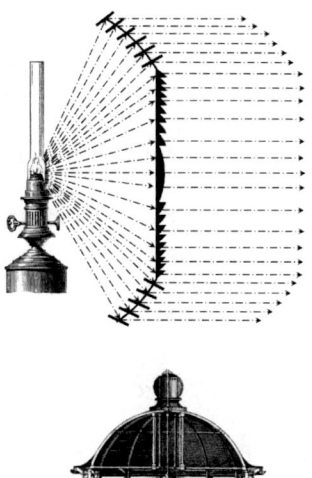

Fig. 4.16 Adolf Ganot's
depiction of a lighthouse
using the Fresnel's
composite catadioptric lens

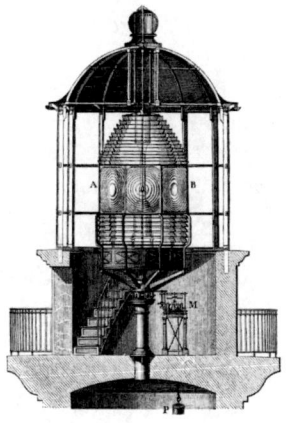

Fresnel's work is concentrated in the three volumes of the *Œuvres Complètes* (Fresnel, 1866, 1868, 1870) and in the 1900 collection of wave-light-theorists edited by Huygens et al. (1900). In his address to the French Academy of Sciences in 1815, title *La Diffraction de la Lumière*, he examined the phenomenon of the coloured fringes by the shadows of bodies illuminated by a luminous point source (Fresnel, 1866), Fresnel explained the phenomenon in purely wave-theoretic manner, through the interference of spherical waves from point sources. This way, he explained the phenomenon of diffraction and from that starting point, he argued that also reflection and refraction could be explained by adopting the wave theory. He proved that, although the light is reflected and refracted in infinite directions, only straight-line directions are actually seen, as interference cancels all other directions. "C'est que leurs vibrations se contrarient, comme il est facile de le prouver", he stated, meaning, the vibrations are opposed and it is easy to show. Fresnel studied in detail the phenomena of coloured fringes in the shadow of illuminated bodies and concluded that even under purely white light the diffraction of the light waves is the cause of the appearance of colours.

> Les franges, dans la lumière blanche, sont la réunion des ban-
> des obscures et brillantes produites par toutes les espèces d'ondes
> lumineuses dont elle se compose. La largeur de ces bandes étant
> proportionnelle à la longueur d'ondulation varie avec elle; en
> sorte que les bandes obscures et brillantes de diverses couleurs,
> au lieu de se su-perposer parfaitement, empiètent les unes sur les
> autres; d'où résultent des mélanges dans d'autres proportions que
> celles qui constituent la lumière blanche, et par conséquent un
> phénomène de coloration.

According to Fresnel, light is definitely propagating as a transverse wave and this is fundamental to his theory. As expected, he was particularly involved in phenomena of diffraction, interference and polarisation because those phenomena could not be explained by Newton's particle theory of light. In some of the *memoirs* in the *Œuvres Complètes*, Fresnel extended Huygens' theory and provided analytical solutions to the phenomena of diffraction and reflection to support *Huygens' principle* of secondary waves and Young's theory of interference.

Further, Fresnel was deeply concerned with *polarisation of light* (see for example the second section of Volume I of the *Œuvres Complètes*) and designed a number of interesting experiments with double refraction in crystals, on which he found out that there is no other reason than the occurrence of polarisation in perpendicular planes for the theory and the observation to be in agreement. Apparently, he imposed polarisation by reflection. Furthermore, he provided an explanation (which Thomas Young failed to do) on the chromatic polarisation, again experimenting with crystals.[17]

> L'explication que je viens de donner des phénomènes de sim-
> ple dépolarisation est fondée sur la supposition que la lumière
> polarisée est divisée par une réflexion complète en deux systèmes
> d'ondes polarisés l'un parallèlement, l'autre perpendiculairement
> au plan d'incidence, et séparés par un intervalle d'un huitième
> d'ondulation.

In the third section of the *Théorie de la Lumière*, which is in the second Volume of the *Œuvres Complètes*, published in 1868 (Fresnel, 1868), Fresnel exposed his complete wave theory of light, including explanations of most important phenomena of light. He provided his own estimates of the wavelengths of light that correspond

---

[17] It is generally accepted that Fresnel was the person who coined the terms for linear, circular and elliptical polarisation.

Fig. 4.17 The optics and math of the meniscus element in a Fresnel lens

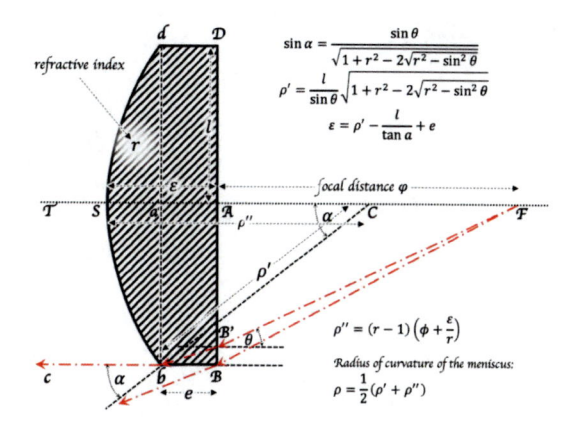

Fig. 4.18 The optics and math of the annular element in a Fresnel lens

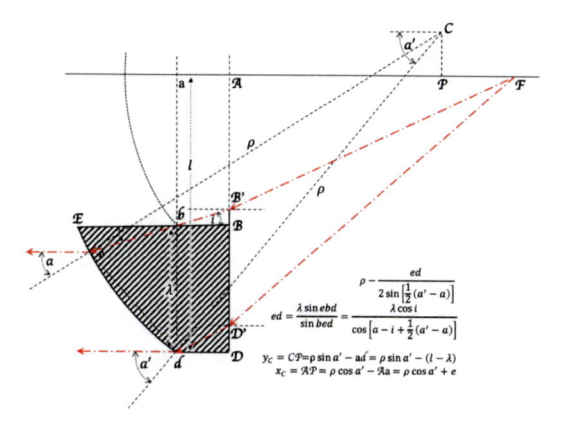

to the principle colours, adopting Newton's set, and extended the estimates to intermediate colours. Although he was involved in every aspect of research targeting the nature and propagation of light, he (like Huygens and contrary to Newton) was not deeply concerned with how colours are perceived by humans, and what colours actually are.

The third Volume of the *Œuvres Complètes*, published in 1870 (Fresnel, 1870), is mostly related to Fresnel's work on lighthouses (*Phares et Appareils d'Éclairage*). This is where one may find all the details of Fresnel's ingenious *catadioptric lenses*. Figure 4.17 is an adaptation to Fresnel's original drawing of the meniscus element of the composite lens. The principle of operation and the required mathematics for the implementation of this element are also included in the figure. Similarly, Fig. 4.18 shows the principle and design data for the annular elements, Fig. 4.19 shows the principle and design data for the catadioptric elements and Fig. 4.20 shows the principle and design data for the catoptric elements of the system.

Fig. 4.19 The optics and math of the catadioptric element in a Fresnel lens

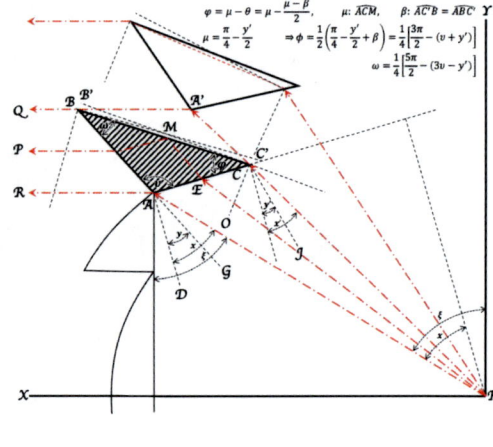

Fig. 4.20 The optics and math of the catoptric element in a Fresnel lens

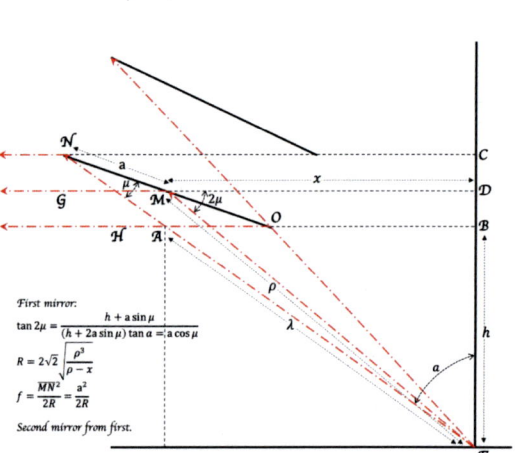

4.11   Arthur Schopenhauer

Arthur Schopenhauer (1788–1860) was a German philosopher, born in Danzig (Gdańsk), Polish–Lithuanian Commonwealth (Baltic coast of Northern Poland). He built on Kant's transcendental idealism and was concerned with the phenomenal world, which he show through a philosophical pessimism, and he has been largely influential for subsequent thinkers. Schopenhauer's most famous work of 1818 *Die Welt als Wille und Vorstellung* (The World as Will and Representation[18]) Schopenhauer (1969a, b) begins with

---

[18] The digitised original Gernal text can be found @ https://www.lernhelfer.de/sites/default/files/lexicon/pdf/BWS-DEU2-0958-03.pdf.

> "Die Welt ist meine Vorstellung:" - dies ist die Wahrheit, welche in Beziehung auf jedes lebende und erkennende Wesen gilt...Die Welt ist Vorstellung.

The world is my conception, says Schopenhauer and this is a truth valid with reference to every living and knowing being. The world is representation. The text continues with

> ...sondern zu welcher nur tiefere Forschung, schwierigere Abstraktion, Trennung des Verschiedenen und Vereinigung des Identischen führen kann, - durch eine Wahrheit, welche sehr ernst und Jedem, wo nicht furchtbar, doch bedenklich seyn muß, nämlich diese, daß eben auch er sagen kann und sagen muß: "Die Welt ist mein Wille".

To Schopenhauer only deeper investigation, more difficult abstraction, the separation of what is different, and the combination of what is identical can lead us to this truth; this truth, which must be very serious and grave if not terrible to everyone, is that a man also can say and must say that "the world is my will". One may probably detect some philosophical similarities to Protagoras' theory. Apparently, Schopenhauer describes as *will* that which characterises the inner nature of all things, and the world seems to consist of two sides, the *world is will* and *the world is representation*. As will, the world is in itself, a unity. As representation, the world is that of appearances, of ideas or of objects, a diversity. What he talks about, is the world as *Reality* and the world as *Appearance*.

As expected, Schopenhauer was deeply concerned with vision and colours and wrote a treatise on this subject titled *Über das Sehn und die Farben* (*On vision and colours*)[19] (Schopenhauer, 1816). In Schopenhauer & Runge (2010) there is a 2010 English translation that also includes the treatise *Farben-Kugel* (*Color sphere*) by Philipp Otto Runge (1777–1810), a Romantic German painter, who derived a trichromatic colour model in the form of a sphere around 1807 (Runge, 1810) and shared it with Goethe.[20] As can be easily deduced from Schopenhauer's treatise, he was among the very first to propose a model for visual perception that separates sensation and representation. In this model, *the role of perception is to transform the subjective sensations of the objects of the outside world into objective representations within, through the interference of the understanding.*

In the part of the treatise devoted to vision, Schopenhauer praises the value of vision among the senses, but, at the same time, clarifies that its value is confined only

---

[19] The original 1816 text may be found online @ https://archive.org/details/bub_gb_qIw5AAAAcAAJ/, or @ https://upload.wikimedia.org/wikipedia/commons/f/f0/Ueber_das_Sehn_und_die_Farben.pdf.

[20] This English translation of the two treatises included in *On vision and colours* can be found online @ https://pdfget.com/pdf-epub-on-vision-and-colors-on-vision-and-colors-color-sphere-download/.

within acquiring a sensation, not perception. He argues that if the eye was the organ of perception, then the world would be perceived inverted as the image formation on the retina follows the laws of optics, thus the eye should only be thought of as the object of sensation. Schopenhauer states that

...sondern die Anschauung entsteht dadurch, daß der Ver-stand den auf der Retina empfundenen Eindruck augen-blicklich auf seine Ursache bezieht, welche nun eben dadu-rch sich im Raum , seiner ihn begleitenden Anschauungs form, als Objekt darstellt. Bei diesem Zurückgehn nun von der Wirfung auf die Ursache, verfolgt er die Richtung, welche die Empfindung der Lichtstrahlen mit sich bringt; wodurch wieder alles an seine richtige Stelle kommt, indem jetzt am Objekt sich als oben darstellt, was in der Empfin-dung unten war.

The perception arises from the fact that the understanding instantly relates the impression felt on the retina to its cause, which now presents itself as an object in space, its accompanying form of perception. Following the reverse path from the effect to the cause, the understanding follows the direction that the sensation of the rays of light brings with it, whereby everything comes back into its right place, and the sensation of bottom corresponds to the top of the object. Schopenhauer is also among the first to provide a definition of illusions in contrast to errors, stating that *an illusion is a deception of the understanding, thus opposed to reality, whereas an error is a deception of reason, thus opposed to truth.*

In the part of the treatise devoted to colour, Schopenhauer unfolds his theory of colour perception, by making it clear that most previous scholars were wrong in trying to infer theories about colour by only studying the effects of light when the research should have been focused on the physiological dimensions of the sensation itself (the eye). He emphatically disregarded Newton's theory on this basis, by stating that

Daß er dabei die Siebenzahl einzig und allein der Tonleiter zuliebe gewählt hat, ist nicht dem mindesten Zweifel unterworfen: er durfte ja nur die Uugen aufmachen, un zu sehn, daß im prismatis-chen Spektrum durchaus nicht jieben Farben sind, sondern bloß vier, von denen, bei größerer Entfernung des Prismas, die zwei mittleren, Blau und Gelb, übereinander greifen und dadurch Grün bilden. Daß noch jetzt die Optiker sieben Farben im Speftrum aufzählen, ist der Gipfel der Lächerlichkeit. Wollte man es aber ernsthaft nehmen, so wäre man, 44 Jahre nach dem Auftreten der Goethejchen Farbenlehre, berechtigt, es eine unverschämte Lüge zu nennen: denn man hat nach gerade Geduld genug gehabt.

Fig. 4.21 Reproduction of
Runge's colour space
(sphere)

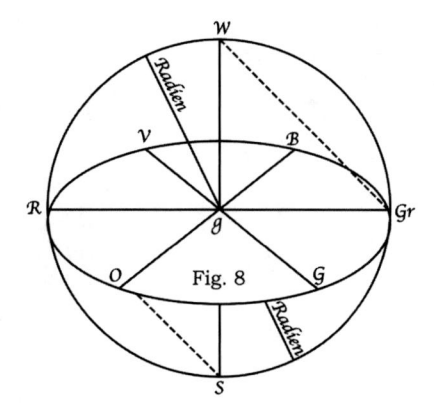

Apparently, he felt that believing Newton's theory in that there are seven primary
colours is totally absurd.

In the development of his theory of colour vision, he stated that when a body
under the influence of light does not react on the eye at all, it should be called black,
whereas white is the light that reacts (the effect of light and white is the same).
Thus the degree of the activity of the retinal reaction indicates the intensity of the
activity, forming thus the grey hues. Furthermore, he defined yellow and violet as
those produced by a (non-equally) divided retina half-activation, complementing
each other. Schopenhauer accepted Runge's colour system, which, in essence, is a
three-dimensional coordinate system with hue, saturation and brightness being the
three coordinates, represented as a sphere, in which the poles are white and black,
pure colours rest on the equator and grey colours run along the white-black axis. A
reproduction of Runge's colour sphere is shown in Fig. 4.21[21] and Fig. 4.22[22] from
Runge (1810).[23]

Regarding the perception of colour, Schopenhauer stated that colour should be
considered as *the qualitatively divided activity of the retina*. In his colour system,
the number of colours is infinite, but any two opposite colours (like yellow-violet)
contain the full potential of all the others, like in an opponent colour theory. Thus, it
is not correct to refer to individual colours, but only of colour pairs, in a sense that
each pair represents the totality of the activity of the retina divided into two (qualitative) halves. In this colour system, there are *six primary colours, violet, blue, green,
red, orange and yellow*,[24] whereas there are only *three basic chemical colours, blue,
red and yellow* (most probably accepted from Runge's colour system). In addition,

---

[21] Adapted from https://archive.org/details/farbenkugeloderc00rung.

[22] Adapted   from   https://archive.org/details/farbenkugeloderc00rung   and   https://commons.
wikimedia.org/wiki/File:Runge_Farbenkugel.jpg.

[23] The original text can be found online @ https://archive.org/details/farbenkugeloderc00rung.

[24] Schopenhauer thinks of the primary colours as excitations of the retina and assigns zero (0) to
black and one (1) to white, which he does not include in the set of primary colours, but he uses
them only as the limits of the excitation range. In this representation, the colours in the set {violet,
blue, green, red, orange, yellow} correspond to a set of activation intensities as {1/4, 1/3, 1/2, 1/2,
2/3, 3/4} respectively.

Fig. 4.22 Various
representations of Runge's
colour space

the system of primary colours is additive, meaning that combined excitations result in summing their individual contributions such that

$$
\begin{aligned}
\mathcal{RED} &= \text{the full activity of the retina minus } \mathcal{GREEN} \\
\mathcal{GREEN} &= \text{the full activity of the retina minus } \mathcal{RED} \\
\mathcal{RED} + \mathcal{GREEN} &= \text{the full activity of the retina} = \\
&= \text{the effect of light (or white)}
\end{aligned}
\tag{4.1}
$$

4.12   Johannes Müller

Johannes Peter Müller (1801–1858) was born in Koblenz, Rhin-et-Moselle, First French Republic (later part of Germany). He was a German physiologist, anatomist, ichthyologist, and herpetologist. His inquisitive mind and influence of other scholars led him, on one hand, to dismiss the approaches to the physiology of his times and, on the other hand, to express his theories in his most important work, the *Handbuch der Physiologie des Menschen*, which was published between 1833–1840 (Müller, 1835; J. Müller, 1838; Müller, 1840b) and was later translated to English as *Elements of Physiology* by William Baly, between 1837–1843 (Müller, 1840a; J. P. Müller, 1842). This publication, which became the textbook in physiology, signified the beginning of a new era in physiology, bringing together anatomy, chemistry and microscopy.

Among the most important contributions is Müllers work on the mechanisms of the nervous system and the senses. He recognised how the sense organs are fit to a single sensation, so distinctively that any type of stimulus would always be perceived in the way the sense organ reacts to it (for example mechanical stimulation of the retina will result in the perception of images, exactly like when light excites it). He conducted a massive amount of experiments to prove this idea.

Müller, in Volume II of the Handbook, expressed some important laws that are of particular interest in relation to the topics of this treatise. He stated that *humans cannot be directly aware of objects in perception, but only of the qualities specific to particular nerves.* Sensations correspond in their features mainly to states in the nerves induced by stimulation and not to states of their distal causes. The characters of sensations are tied specifically to the nerves producing them and to the 'energies' (in quotes here, although this is the expression Müller used) of those nerves. Sensations need not have external causes at all but can be caused internally by direct stimulation of a nerve or even by an electrical impulse.

His general contribution to the physiology of the nerves can be found in Book III of Volume I of the Handbook, where he describes the structure and excitability of the nerves and the propagation of signals and actions.

Book V. (hosted in Volume II) Müller begins with his preliminary considerations *Of the Senses*, stating that

> Die Sinne unterrichten uns von den Zuständen unseres Körpers durch die eigenthümliche Empfindung der Sinnesnerven, sie inderrichten uns auch von den Eigenschafted und Veränderungen der Natur ausser uns, insofern diese Zustände unserer Sinnesnerven hervorrufen. Die Empfindung ist allen Sinnen gemein, aber der modus der Empfindung ist in den enzelnen verselieden, nämlich Lightempfindung, Tonempfindung, Geschhnnack, Geruch, Gedühl.

As far as qualities and changes in internal states and the external nature give rise to changes in the conditions of nerves, senses arise to inform of those changes. Sensation is common to all senses, but the type of sensation is different in each sense, thus differentiating the sensations of light, sound, taste, smell, and of feeling or touch. Furthermore, paragraph V begins by stating that

> Die Sinnesempfindung ist nicht die Leitung einer Qualität oder eines Zustandes der äusseren Körper zum Bewustsein, sondern die Leitung einer Qualität, eines Zustandes eines Sinnesnerven zum Bewustsein, verunlasst durch eine äussere Ursache, und diese Qualitäten sind in den verschiedenen Sinnesnerven verschieden, die Sinnesenergieen.

A bold statement indeed, conveying that *sensory perception is not the conduction of quality or a state of an external body to consciousness, but the conduction of a quality or a state of the corresponding sensory nerve, caused by an external cause.* Clearly, perception gives an interpretation of the natural world by coupling stimuli from it with knowledge of the mechanics of the corresponding sensor.

The entire first section of Book V is devoted to sight (*I.Abschnitt. Vom Gesichtssinn—Of Vision*). First, Müller analyses the physical conditions necessary for the formation of luminous images. He then provides an excellent description of the structure and operation of the eye (a) by means of a pinhole camera and (b) by means of a focusing system based on a refractive lens. Thus, he states the basic laws of reflection and refraction to provide a basis for the description of the mechanisms of the vision. He also mentions aberrations of lenses. For Müller the lens is the most perfect part of the eye, having the shape and transparency for the required refraction to focus images on the retina.

> Körper, welche das Licht in jenem Sinne zu sammeln vermögen, sind die durchsichtigen das Licht brechenden Mittel, deren vollkommenste für das Sehorgan zweckmässigste Gestalt die linsenförmige ist, wie sich specieller sogleich ergeben wird.

Müller accepted Newton's colour theory and cited the effects on prisms. He reiterated the conclusion that white light is composed of the different coloured rays, which combined give the sensation of the white, but which may be separated by refractive media, due to their different refrangibility.

> Diess führt zu dem Schluss, dass das Weisse dann gesehen werde, wenn dieselben Stellen eines Körpers ungleichartige Strahlen aller Art zugleich erhalten und ins Auge werfen, das hingegen die Farbe dann erscheine, wenn das gleichartige Licht einer Art den Eindruck hervorbringt, mit anderen Worten, dass das weisse Licht aus den verschiedenen Farben zusammengesetzt sei, welche zusammen weiss geben, durch brechende Mittel abet wegen ihrer verschiedenen Brechbarkeit zur Sonderung gebracht werden.

Nevertheless, In subsequent paragraphs, he dismissed Newton's seven dioptric colours theory and proposes that there should only be three primary colours, *yellow, blue, and red*. He defined complementary those colours that their combinations produce white. He was careful to denote in this definition that the one should be a homogeneous colour and the other a mixed prismatic colour, thus discussing combinations like green and red, or violet and yellow, blue and orange. In addition, he recognised that black colour is by definition the absolute darkness, the state of repose or freedom from excitement. Figure 4.23 is a reproduction of the colour circle Müller provided in this Handbook, in which he put the primary colours on the vertices of an equilateral triangle, circumscribed a circle on that triangle, and placed the composite colours between the primary colours that mix to produce them. Diameters in this circle denote the complementary colours, as he defined them.

Fig. 4.23 Müller's colour circle

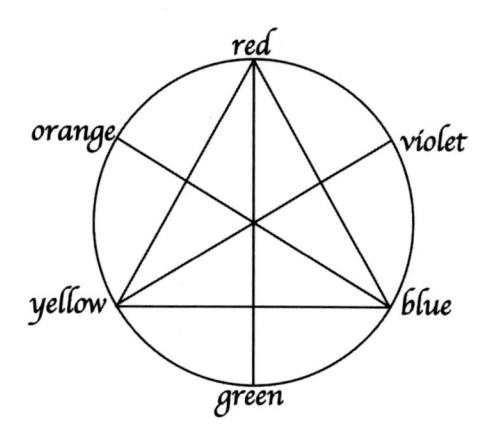

Table 4.2 Hershel's measurements of common colours

Colours of the Spectrum	Length of an undulation in millionths of an inch	Number of undulations in an inch	Number of undulations in a second (trillion)
Extreme red	26.6	37640	458
Red	25.6	39180	477
Intermediate	24.6	40720	495
Orange	24.0	41610	506
Intermediate	23.5	42510	517
Yellow	22.7	44000	535
Intermediate	21.9	45600	555
Green	21.1	47460	577
Intermediate	20.3	49320	600
Blue	19.6	51110	622
Intermediate	18.9	52910	644
Indigo	18.5	54070	658
Intermediate	18.1	55240	672
Violet	17.4	57490	699
Extreme violet	16.7	59750	727

Müller dismissed Goethe's objections to Newton's theory of colours as erroneous, by stating that there was already proof of how light interacts with translucent media causing the appearance of a particular colour, which was the main source of objection in Goethe's approach. In conclusion, he reiterated John Frederick William Herschel's (1792–1871) measurements on the wavelengths, wavenumbers and frequencies of selected colours (as found in the updated English translation), as shown in Table 4.2.

In *II.Capitel. Vom Auge als optisehem Werkzeuge*, Müller analysed the eye as the optical instrument of vision, where he distinguished three types of eyes based on the complexity of their structure and provided a description for these types. The

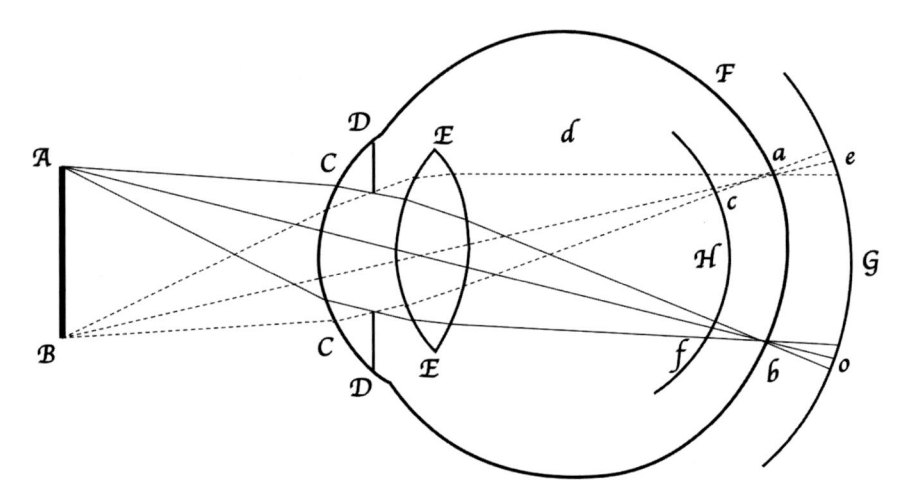

Fig. 4.24 Müller's diagram of the refractions in the human eye

human eye, which is in the category of those based on refractive lenses, is described in a manner relating to the function of vision. He provided a brief description of the basic parts of the eye and particularly focused on the retina, which he described as consisting of three main layers and focused on the layer of the cone or rod-shaped nerve terminations. A reproduction of Müller's diagram of the refractions taking place in the human eye is shown in Fig. 4.24. As he stated, the refraction of the rays of light in the eyes is threefold, one by the cornea and two by the lens, as shown by the bending of the rays in the figure. He recognised that the focal distance of the eye should be exactly on the retina, otherwise any image would not be perfectly focused and would appear blurry, as shown in the figure by the two spherical surfaces H and G, in which the retina is positioned closer and further to the lens.

Müller included in his *Handbuch* all the known knowledge of that time regarding the structure and function of the eye as a focusing apparatus and suggested that any image is a composition of the tiny sensations captured by the rod-like optic nerve terminations in the eye. He devoted a part of the text in the description of the adaptation at different distances, which he suggested (as Young did) could be attributed to a change in the convexity of the lens, and not the enlargement of the eye, although other experts also suggested the change in cornea's convexity. He further discussed the issues of *chromatic aberration*, a phenomenon that results from the dispersion property of light, by which different wavelengths undergo different refraction. In the case of the human eye, which has a lens of a fixed refractivity, this results in having different focal lengths for lights of different wavelengths, thus short wavelengths focus in front of long wavelengths as they are more refrangible. In colour perception, this translates to having violet light focus in front of yellow, which in turn, focuses in front of red. Müller analyses the achromatic property of the eye, the property by which the refractive media of the eye do not disperse the light entering the eye, and supports that the chemical composition and geometric structure makes it possible.

> Worin die Achromasie ihren Grund hat, lässt sich mit Bestimmtheit nicht angeben, wohl aber die Möglichkeit der Achromasie des Auges aus dem optischen Bau desselben einsehen. Seine brechenden Mittel sind von ungleicher Brechkraft, von ungleichen Convexitäten und ungleicher chemischer Constitution. Das eine ist die Linse mit ungleichen Convexitäten, das zweite die Cornea mit dem Humor aqueus. Letztere bilden zusammen eine convex-concave Linse, deren Brechkraft von der Linse verschieden ist. Vielleicht ist die Farbenzerstreuungskraft beider brechender Mittel ihrer Brechkraft nicht proportional und hierdurch die Achromasie bedingt.

In *Chapter III. Of the action of the retina, optic nerve, and sensorium in vision*, Müller reported the important findings of his time regarding the actual chromatic visual perception. He stated that "light and colour are actions of the retina, and of its nervous prolongations to the brain. The kind of colour and luminous image perceived depends on the kind of external impression." Simultaneous impression of undulations of different wavelengths on the same point of the retina results in the sensation of white. Colours are just manifestations of excitation with light of different wavelengths. Müller further outlined the limits of the current knowledge during his time by stating that *there was no clue on where perception takes place.*

> Wo wird der Zustand der Nervenhaut empfunden, in der Nervenhaut selbst oder im Gehirn?

He elaborated on this issue by analysing the facts known at that time; it was already known that the central vision is acute whereas the peripheral was not and that the optic nerve connecting with the brain consists of far fewer fibres than those on the retina. He examined various possibilities but ultimately concluded that at that time it was impossible to form a concrete theory. By analysing how the eye-brain interaction occurs, according to experience and experiments, he suggested that their cooperation is so tight and constant that it is difficult to distinguish what actually influences a visual sensation. For the relation of the sense of vision to the perception of the external world, Müller reminds that *it is by the operation of the judgment that the objects of vision are recognised as exterior to the body of the observer.* In his analysis, he presented several interesting facts regarding visual perception, the results of the simultaneous action of the two eyes (stereo vision) and even discussed the aesthetics of colour combinations.

An interesting essay by Scott Edgar published in 2015 took a more philosophical look at Müller's work. Edgar pointed out that (for the Neo-Kantians, second half of 19th century) the crucial insight from this work from an epistemological perspective is that if the nature of any representation (say of a physical world object) is not subject to the independent properties of the represented object but rather to the properties of the sensory and cognitive system, then the representation will not resemble the independent object (Edgar, 2015).

4.13   William Hamilton

William Rowan Hamilton (1805–1865) was born in Dublin, Ireland. He was a mathematician and astronomer and left important work in pure mathematics and mathematics for physics. He was the founder of a theory of dynamical systems in classical mechanics (later known as *Hamiltonian mechanics*), which became crucial to the study of electromagnetism and the development of quantum mechanics. He is also known for his invention of the *quaternions*. Quaternions form a generalisation of the theory of complex numbers in the case of four dimensions, to describe rotations in three dimensions. As complex numbers form a two-dimensional space to describe one-dimensional vibrations (as projections of 2D on 1D), quaternions form a four-dimensional space to describe three-dimensional rotation (as projections of 4D on 3D). Although vector analysis replaced quaternions the following centuries, the compact and quicker computations they involve in comparison to matrix representations, make quaternions extremely useful (and stable) in practical applications involving three-dimensional rotations (aviation, computer graphics, etc.). As extensions to complex numbers, quaternions are similarly represented, for real numbers a, b, c, d and unit quaternions $\mathbf{i}, \mathbf{j}, \mathbf{k}$ as

$$
\begin{aligned}
& a + b\mathbf{i} + c\mathbf{j} + d\mathbf{k} \\
& \mathbf{i}^2 = \mathbf{j}^2 = \mathbf{k}^2 = \mathbf{i}\mathbf{j}\mathbf{k} = -1
\end{aligned}
\tag{4.2}
$$

Quaternions, although a powerful and straightforward tool to use, are very difficult to grasp, due to their four-dimensional nature, and a number of treatises and resources can be found that deal with their description, analysis and representation.[25]

At the end of the 1820s, Hamilton presented a theory of a single function, *Hamilton's principal function* (also known as the *Hamilton–Jacobi equation*), which bridges mechanics, optics, and mathematics.

$$
S(q, t; q_0, t_0) = \int_{t_0}^{t} \mathcal{L}\left(\gamma(\tau), \dot{\gamma}(\tau), \tau\right) d\tau
\tag{4.3}
$$

where t is time, q is a position in generalised coordinates, (t_0, q_0) are initial conditions, $\dot{q}$ is the velocity, γ are the solutions to the Euler-Lagrange equations (extrema), $\gamma|_{\tau=t_0} = q_0$ and $\exists \hat{t} \in [t_0, t_1) \; \gamma|_{\tau=\hat{t}} = q$. *This is a milestone in mechan-*

---

[25] An interesting explanation, demonstration and interactive simulation can be found online @ https://eater.net/quaternions.

Fig. 4.25 Hamilton's wave
propagation of light

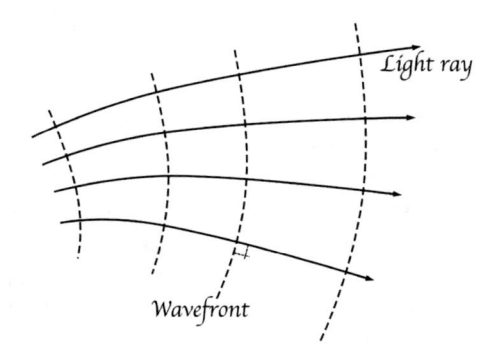

*ics, since it is the only formulation of mechanics, in which the motion of a particle
can be represented as a wave.* This gave rise to Hamilton's representations of light
propagation as shown in the diagram of Fig. 4.25.

Hamilton viewed all processes, thus also light propagation, in this way, and based
his theory of light, which he published as *Theory of Systems of Rays* in 1828, on such
an approach (Hamilton, 1828a, b). There he focused on reflection and refraction, the
most basic phenomena of light propagation studied by all scientists in optics. He
presented the ellipsoid model to his reflection theory, by which any point on the ray
of incidence and the corresponding point of reflection are foci of an ellipsoid that
touches the surface of reflection.[26]

Hamilton formulated the principle of stationary action (or principle of least
action) for dynamical systems. Of course, this principle appeared many times in
history (of the science of the optics), even during Euclid's and Hero's era and the
formulation for light is credited to Pierre de Fermat in the 1600 s. Hamilton's principle generalises the case for the dynamics of physical systems. According to this
principle, *the path taken by a system between two times and two configurations is
the one for which the action is stationary (no change) to first order.* Given two
states of a system expressed in N-generalised coordinates $q_i(t)$, $(i = 1, 2)$ that correspond to two times t_1, t_2, the true evolution of the system is a stationary point of
the action-describing functional,

$$S[\mathbf{q}] = \int_{t_1}^{t_2} \mathcal{L}(t, \mathbf{q}(t), \dot{\mathbf{q}}(t)) \, dt \tag{4.4}$$

where $\mathcal{L}(t, \mathbf{q}, \dot{\mathbf{q}})$ is the Lagrangian of the system. By this definition, S is clearly
a functional, for which Hamilton's principle states that the true evolution of the
system is a solution of the functional

---

[26] This is due to the *reflective* nature of ellipses, by which when connecting any point p on an
ellipse to its foci f_1, f_2 the tangent to that point forms a surface of reflection for one of the 'radii'
(say $|p\,f_1|$) to the other (say $|p\,f_2|$). Due to this reflective property of ellipses, all possible rays
passing through a focal point are expected to be reflected by the ellipse towards a direction that
necessarily passes through the other focal point.

$$\frac{\delta S}{\delta \mathbf{q}(t)} = 0 \tag{4.5}$$

which gives the path, in configuration space, for which the action is stationary.

Hamilton's work on optics is complemented with his three important supplements to the *Theory of Systems of Rays*, sequentially published during 1830 and 1831 (Hamilton, 1830b, c, a, 1831a, b, 1837). In these supplements, he extended his theory and made it clear that he supported a *wave theory of light*. One of the remarkable results of his work is the prediction of *conical refraction*, a phenomenon occurring in biaxial crystals, by which light rays exit those materials in the form of a hollow cone. Although Hamilton left a voluminous work on optics, he was not particularly concerned about the way humans perceive and interpret colours.

4.14 Hermann Günter Grassmann

Hermann Günter Grassmann (1809–1877) was a German polymath, linguist, mathematician and physicist, born in Stettin, Kingdom of Prussia (Western Poland). His legacy includes contributions to linear algebra, projective and differential geometry, but also linguistics. It should be noted that he is probably the first to apply vector methods to mechanics and was the first to formulate a theory of linear algebra, so ahead of his time, that it initially did not attract any attention as it was not understood. In 1853 he published *Zür Theorie der Farbenmischung* (*Theory of Compound Colours*), his theory on colour mixing and colour sensation, known today as *Grassmann's laws* in optics (Grassmann, 1853, 1854; MacAdam, 1970). Grassmann's theory plainly states that that chromatic sensation in human vision can be described as an effective stimulus consisting of linear combinations of different wavelength lights. His *first law* states that colour matches are trivariate; provided three colour primaries (say, R, G, B), any colour C can be matched by a weighted summation of the three primaries (measured in any form of quantity of light power r, g, b).

$$C \equiv r(R) + g(G) + b(B) \tag{4.6}$$

Grassmann's *second law* states that mixing any two colours is matched by a linear combining of the mixtures of any three other colours that individually match the two colours considered.

$$
\begin{aligned}
C &\equiv C_1 + C_2 \\
C_i &\equiv R_i + G_i + B_i, \quad i = 1, 2
\end{aligned} \tag{4.7}
$$

Grassmann's *third law* states that the hue of a colour resulting from additive colour mixing depends only on the colour impression of the initial colours, but not on their physical (spectral) compositions. Thus mixing of even the metameric colours[27] can be described exactly on the basis of their colour impression, and conversely, no direct conclusions about the spectral composition of colour can be drawn from the mixing.

$$C' \equiv k \cdot C \equiv k \cdot R + k \cdot G + k \cdot B \tag{4.8}$$

Grassmann's *fourth law* states that the intensity of an (additively) mixed colour corresponds to the sum of the intensities of the mixed colours.

$$I(C_3) \equiv I(C_1) + I(C_2) \tag{4.9}$$

Of particular interest is that Grassmann opens his 1853 paper (Grassmann, 1853) by clearly confronting Helmholtz, stating that

Im 87. Bande dieses Journals theilt Hr. Helmholtz eine Reihe zum Theil neuer und sinnreicher Beobachtungen mit, aus welchen er den Schlufs zieht, dass die seit Newton allgemein angenommene Theorie der Farbenmischung in den wesentlichsten Punkten irrig sey, und es namentlich nur zwei prismatische Farben gebe, nämich Gelb und Indigo, welche vermischt Weiss liefern.

Hierbei wird es nöthig seyn, den Farbeneindruck, dessen das Auge fähig ist, in seine Momente zu zerlegeg. Zunächst unterscheidet das Auge farbloses und farbiges Licht. An dem farblosen Lichte (Weiss, Grau) unterscheidet es nur die größere oder geringere Intensitaet, und diese lässt sich mathematisch bestimmen. Ebenso unterscbeiden wir an einer homogenen Farbe nur ihre größere oder geringere Intensität. Aber auch fuer die Verschiedenheit der einzelnen homogenen Farben haben wir ein mathematisch bestimmbares Maaß, welches uns am vollkommensten in der jeder Farbe entsprechenden Schwingungsdauer geboten wird; schon die populäre Sprache hat diese Differenz auf eine sehr passende Weise durch den Ausdruck Farbenton bezeichnet. Wir werden also an einer homogenen Farbe zweierlei: ihren Farbenton und ihre Intensität unterscheiden können. Vermischt man nun eine homogene Farbe mit farblosem Lichte, so wird der Farbeneindruck durch diese Beimischung abgeschwächt.

---

[27] Metameric colours are those with the same colour impression but with a different spectral composition.

which may be rendered in English as

In the 87th volume of this journal, Mr.Helmholtz conveys a series of partly new and ingenious observations, from which he concludes that the theory of colour mixing, generally accepted since Newton, is erroneous in the essential points, and in particular only two prismatic colours give, as yellow and indigo, which deliver mixed white.

Here it will be necessary to dissect the colour impression of which the eye is capable, into its elements. First, the eye distinguishes colourless and coloured light. At the colourless light (white, grey) it distinguishes only the greater or lesser intensity, and this can be determined mathematically. Likewise, we only consider their greater or lesser intensity in a homogeneous colour. But also for the difference of the individual homogeneous colours, we have a mathematically determinable measure, which is offered to us most completely in the oscillation period corresponding to each colour; even the popular language has designated this difference in a very fitting way by the term colour tone (hue). So we will be able to distinguish two things from a homogeneous colour: its colour tone (hue) and its intensity. If one then mixes a homogeneous colour with colourless light, the impression of colour is weakened by this admixture.

Grassmann claimed that in white light one is able to only distinguish the intensity, whereas in a homogeneous colour both its intensity and its hue can be distinguished. *Every impression of colour may be analysed into three mathematically determinable elements, the hue, the intensity of colour, and the intensity of the intermixed white.* Apparently, this is the basis of any hue-saturation-brightness (or lightness, or intensity) colour model. Quite interestingly, due to his linguist background, Grassmann recognised and acknowledged that it is rather difficult to prove this argument and invoked the usage of the language to his defence, in that there has never been an observer to name other elements (apart from these three) for the description of the impression of colour.

Grassmann supported that to any homogeneous colour one may find another, which, if mixed with the former, would result in colourless light. He carefully analysed and compared the work of Newton and Helmholtz and derived his set of complementary colours and his version of the *colour circle*, based on that of Newton's, as shown in Fig. 4.26. In this diagram, the letters A to G correspond to the *Fraunhofer lines*.[28] Grassmann's colour circle was derived from Newton's circle according

---

[28] The Fraunhofer lines are a set of spectral lines–named after the German physicist Joseph von Fraunhofer (1787–1826)–which were originally observed as absorption lines in the optical spectrum of the sun, observed as a result of gas in the photosphere of the sun. Practically, around 1817 Fraunhofer examined further the observation of chemist and physicist *William Hyde Wollaston* (1766–1828), who noticed gaps in the spectrum of the sun through a prism. By looking even closer, Fraunhofer found a large number of missing slices (or dark lines) in the spectrum. These are

Fig. 4.26 Reproduction of Grassmann's colour circle adapted from Grassmann's original work and MacAdam (1970)

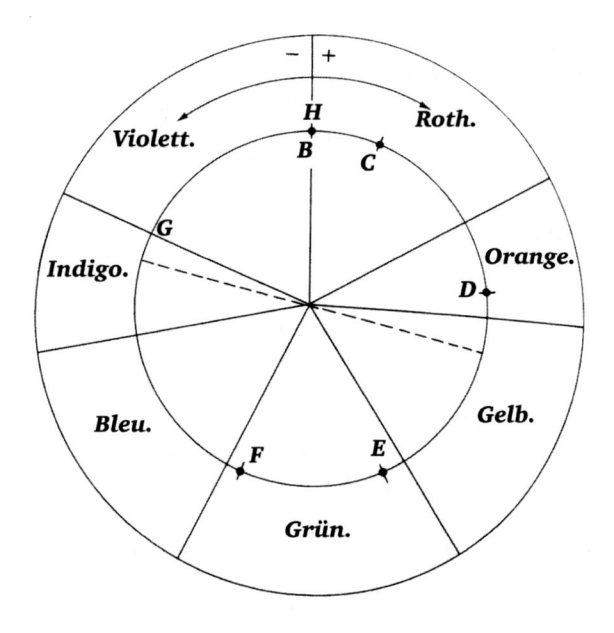

Table 4.3 Grassmann's complementary colours

Yellow	Yellowish Green	Green	Bluish Green	Azure	Indigo
Indigo	Violet	Purple	Red	Orange	Yellow

to the rules in the *Opticks* and the corresponding positions of the Fraunhofer lines, and Grassmann provided detailed relations and proportions among the various distances on the circle. Grassmann's table of complementary colours as outlined in Grassmann (1853) are listed in Table 4.3, where the complementary colours stand one above the other.

4.15 Hermann von Helmholtz

Hermann Ludwig Ferdinand von Helmholtz (1821–1894) was born in Potsdam, Kingdom of Prussia (Germany). He was a physicist, physician and philosopher with a significant scientific contribution in many fields of research, including theories of energy conservation, electrodynamics, thermodynamics, philosophy of science, aes-

---

the lines named after him and are attributed to the absorption of particular wavelengths of light by the various materials. It was around 1860 that *Gustav Robert Kirchhoff* (1824–1887) and *Robert Wilhelm Eberhard Bunsen* (1811–1899) found out why these dark lines appear, by working in the opposite direction, studying the emission lines of various heated gasses. Thus, the dark lines in the solar spectrum seen on the surface of the Earth are due to the absorption of light by the gasses that make up Earth's atmosphere.

thetics and more. Helmholtz was particularly intrigued by the physics of perception and focused on optics and acoustics. He is particularly known for his theories of vision and visual perception. Helmholtz quickly became famous for his 1851 invention of the *ophthalmoscope*, a device with which it is possible to see the retina of the eye (*fundus* image of the eye[29]).

In the 1860s, with the appearance of Helmholtz's *Handbuch der physiologischen Optik* (*Handbook of Physiological Optics*) (von Helmholtz, 1867, 1909a, b, c),[30] old and new experimental observations have been unified and explained under a single theory. This theory has long been known as *the Young-Helmholtz theory of colour vision*, though some researchers argue that it should be known as *the Young-Helmholtz-Maxwell theory of colour vision* (Sherman, 1981; Kremer, 1993; Heesen, 2015). The famous "Fig. 95" from the *Handbook*, a drawing of Helmholtz's ophthalmoscope, is reconstructed in Fig. 4.27 from the 1867 edition (this figure became "Fig. 104" in the 1909 edition of Volume I of the *Handbook*). Figure 4.28 shows Helmholtz's diagrams for the ophthalmoscope apparatus (from the 1909 edition of Volume I of the *Handbook*). The description of the principle of operation by Helmholtz himself is rather simple and straightforward.

Sehr viel bequemer wird die Beobachtung, wenn der Beobachter einen durchbohrten undurchsichtigen Spiegel anwendet, um das Auge $\mathcal{A}$ zu erleuchten. Es sei in Fig. 95 wieder $\mathcal{A}$ das beobachtete, $\mathcal{B}$ das beobachtende Auge, C die Convexlinse, und SS ein durchbohrter Spiegel. Von dem Netzhautpunkte a wird ein Bild bei d entworfen, welches der Beobachter durch die Oeffnung des Spiegels hin betrachtet. Von dem ganzen von a kommenden Strahlenkegel geht nur der schmale Theil für die Beleuchtung verloren, welcher durch die Oeffnung des Spiegels fällt, der ganze übrige Theil wird reflectirt und kann dem leuchtenden Körper zugelenkt werden. Zu dem letzteren Ende ist entweder der Spiegel SS ein Hohlspiegel (Ruete), oder aber ein Planspiegel (Coccius) oder Concavspiegel (Zehender), neben dem man eine Linse $\mathcal{L}$ angebracht hat, welche die Strahlen auf den leuchtenden Körper vereinigt.

Helmholtz describes how to set up an apparatus using a perforated opaque mirror SS before the observer's eye $\mathcal{B}$, in order to look into the eye of the subject $\mathcal{A}$, by illuminating through a source $\mathcal{D}$. Lens $\mathcal{L}$ is used to force the rays of light to hit the mirror in a near parallel arrangement, which then converge to mirror's focal point d, pass through lens C which guides light rays to the correct position into the eye.

---

[29] A *fundus* is a part of a hollow object (particularly an organ) that is furthest from the opening. In the context of vision, a fundus image of the eye shows the retina, the part of the eye that is opposite to the opening, the pupil.

[30] Also in subsequent smaller treatises like *Ueber die Theorie der zusammengesetzten Farben* (*On the Theory of Compound Colours*) and *The recent progress of the theory of vision* (von Helmholtz, 1852b, a; Helmholtz, 1885).

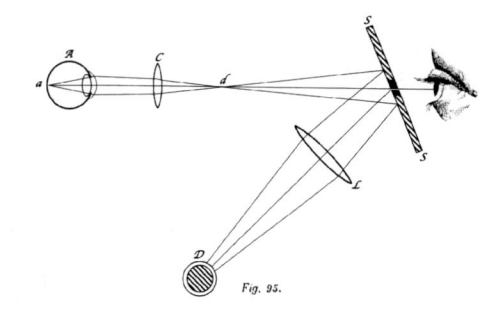

Fig. 4.27 Reconstruction of Helmholtz's drawing of the ophthalmoscope

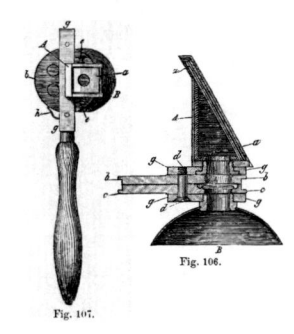

Fig. 4.28 Helmholtz's drawings of the ophthalmoscope apparatus

The reflected light will follow the prescribed path straight to the eye of the observer through the perforated mirror.

In his 1868 lecture on the *Recent progress of the theory of vision* (Helmholtz, 1885), Helmholtz presented an account of the known physiological and mental process of vision. This was a summarisation of the current knowledge, which was finalised, to the extent the technological and theoretical means of his time could support. This is a text worth reading, as it is a view of the structure and physiology of the human eye and its function, which has not changed since then. Figure 4.29 side by side with Fig. 4.30 show how Helmholtz presented the eye as a camera obscura, but with great admiration for the efficiency by which nature created the eye, unparalleled by whatever apparatus human was able to create at that time. The inverted image in the eye formed by refraction was clearly accepted at that time. Furthermore, Fig. 4.31 shows a section of the central part of the retina (part of the *macula*, near *fovea centralis*), the photosensitive part of the eye, which he attributed to the anatomical work of Friedrich Gustav Jakob Henle (1809–1885). According to his description, this is the part responsible for the most significant part of the overall function of vision. Helmholtz informed that it was already estimated that the viewing angle of a single human eye covers roughly 160° laterally and 120° vertically, whereas the two eyes combined cover a horizontal field of view of roughly 180°. He also provided information regarding the visual acuity in central vision, which was estimated to that of the resolution provided by a single cone, roughly in the order of a minute of a degree. Helmholtz recognised *three stages in vision*, beginning with

Fig. 4.29 Helmholtz's depiction of the camera obscura of his time

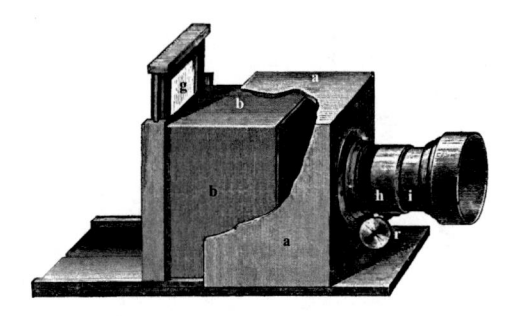

Fig. 4.30 Helmholtz's depiction of the anatomical structure of the human eye

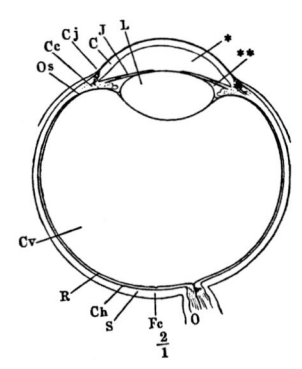

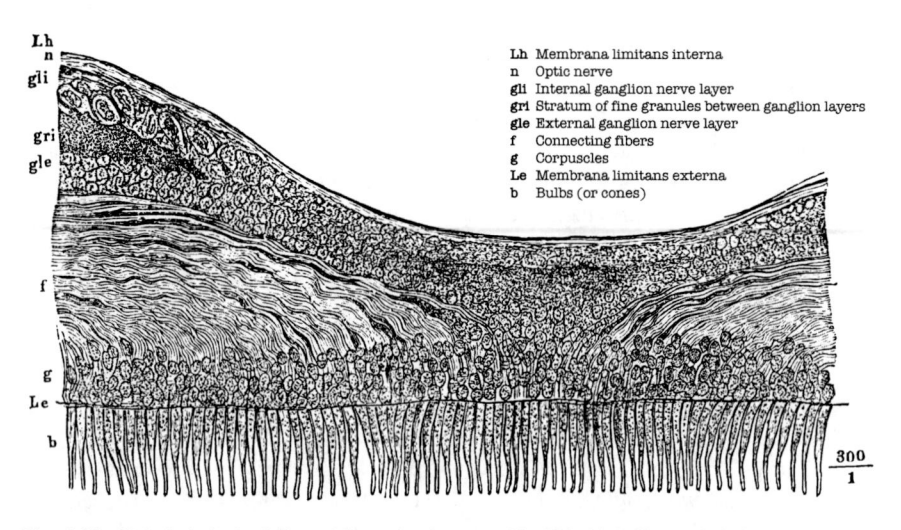

Lh Membrana limitans interna
n   Optic nerve
gli Internal ganglion nerve layer
gri Stratum of fine granules between ganglion layers
gle External ganglion nerve layer
f   Connecting fibers
g   Corpuscles
Le Membrana limitans externa
b   Bulbs (or cones)

Fig. 4.31 Helmholtz's depiction of the retina borrowed by Friedrich Gustav Jakob Henle

the *physical* (the optics of the eye), then the *physiological* (the conversion of light to nervous impulses) and ending with the *psychological* (the perception).

In the *Handbook*, Helmholtz provided an excellent and critical review of the theories of light and colour vision previously developed, going back to the time of Aristotle. He unfolded this review using Newton as a reference point and explicitly mentioned the theory of Goethe and his 'cloudy' media hypothesis, along with Goethe's strong opposition to Newton's theory for the composition of white light. As he stated

> The complexity of white light, which Newton announced, was the first decisive empirical step in the direction of recognising the merely subjective significance of the sense-perceptions. Goethe's presentiment was, therefore, correct when he violently opposed this first advance that threatened to ruin the "fair glory" of the sense-perceptions.

Helmholtz concluded the review with a presentation of the basic elements of the theories of Hegel, Descartes, Hooke, de la Hire, Huygens, Euler, Hartley, Young, Fresnel and Brewster. Then, he proceeded to unfold his theory of the sensation of compound colours by using an analogy and a profound distinction of the visual system with the auditory system for the perception of sound. He emphasised a difference in the 'mechanics' of mixing powdered or liquid pigments to the mixing of lights, denoted an analogy of the light passing through liquids to the light passing through prims, and ultimately defined what today is called the *subtractive colour representation* and a basis for the definition of the effects of optical filters. He emphatically suggested that

> Evidently, therefore, the result of mixing pigments cannot be used to deduce conclusions as to the effect of combining different kinds of light. The statement that yellow and blue make green is perfectly correct in speaking of the mixture of pigments; but it is not true at all as applied to the mixture of these lights.

In Helmholtz's colour theory, white can be produced by combining different pairs of simple colours in definite ratios, which are complementary colours. *He defined the complementary colours of the spectrum to be red and greenish-blue, orange and cyan-blue, yellow and indigo-blue and greenish-yellow and violet.* He defined the complementary colour to green to be purple, which is not a single colour (but rather a compound colour). In his theory, the most saturated colours in order of degree of saturation are violet, indigo-blue, red and cyan-blue, orange and green, yellow. In this view, the number of different colours is exhausted by mixing pairs of two simple or homogeneous colours. Helmholtz's colour representation system for light sources includes three variables, namely *luminosity, hue and saturation*, which may produce every colour impression. In introducing this colour model he rejects a colour model based on primary colours, like the one based on red, yellow and blue, as, he states, no combination of three colours can actually reproduce the effect that the prismatic colours have. Nevertheless, he proceeds and proposes a change in the set of primary colours into violet, green and red, as more suitable for better approximation of the prismatic colours. He defines a new colour circle, based on the intuition that many colours are not represented in triangular regions as

Fig. 4.32 The flattened
colour circle that Helmholtz
proposed

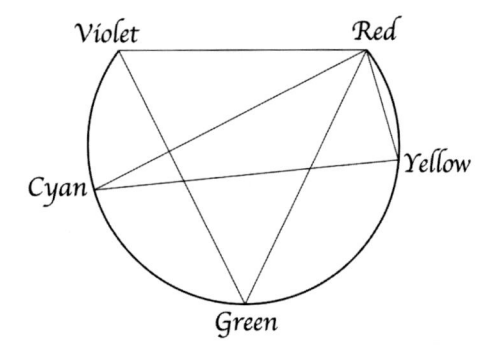

shown in Fig. 4.32. Through this graph, he states that it is clear how a red-yellow-
blue (blue was actually cyan during that time) model is totally inappropriate for
colour representation. Based on experiments (which he does not mention further
at this point) he proposed that the colour circle should actually be flattened at the
violet-red region. In a way, this diagram, if inverted, was very close to the most
modern colour representation theories. Regarding the proper and objective naming
of the colours, Helmholtz provided Table 4.4, in which colour names were matched
to wavelength and corresponding Fraunhofer lines.

Table 4.4 Objective naming of colours by Helmholtz

Fraunhofer lines	Wavelength (nm)	Naming
A	760.40	Extreme red
B	686.853	Red
C	656.314	Border of red and orange
D	589.625 589.023	Golden yellow
E	526.990	Green
F	486.164	Cyan blue
G	430.825	Border of indigo and violet
H	396.879	Border violet
L	381.96	
M	372.62	
N	358.18	
O	344.10	Ultraviolet
P	336.00	
Q	328.63	
R	317.98	
U	294.77	

Fig. 4.33 The spectral sensitivity of the eye's nervous fibres according to Helmholtz

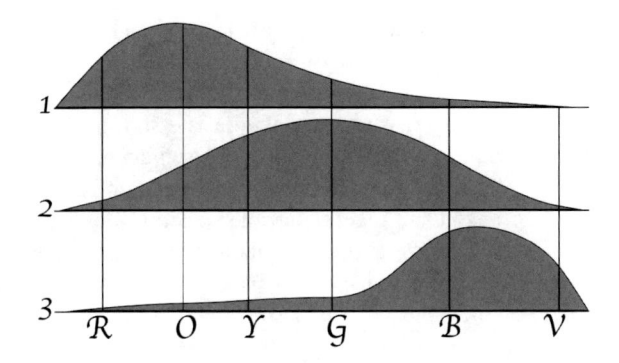

Although Helmholtz extended the space of the colour model to accommodate for more perceived colours, he emphatically pointed out that *no set of primary colours should be taken as having any objective significance* whatsoever, when the human eye is not taken into account. Here, he accepts *Thomas Young's theory of colour sensation*, in which, he agrees, there is the meaning of choosing and talking about sets of primary colours. In this view, the eye consists of *three distinct sets of nervous fibres*, which when excited correspondingly create the sensation of red, green and violet colours. Depending on its wavelength, light excites these fibres accordingly, although, the fibres should not be expected to have a strict wavelength selection mechanism; light should be expected to excite all fibres to a different degree depending on their nature. Although Helmholtz did not provide any evidence whatsoever for a quantitative analysis on this subject, he presented a graph of the response he envisioned these fibres should have in order to create the sensation of colour, as shown in Fig. 4.33. In this figure, the horizontal axis is in decreasing wavelengths from red ($\mathcal{R}$) to violet ($\mathcal{V}$).

Helmholtz, in Sect. 20 *Die zusammengesetzten Farben* of the 1867 edition of the *Handbook*, stated that

Will man dagegen in der Farbentafel als gleich gross solche Quantitäten verschiedenfarbigen Lichts betrachten, welche dem Auge bei einer gewissen absoluten Lichtintensität als gleich hell erscheinen, so erhält die Curve der einfachen Farben eine ganz andere Gestalt ähnlich wie in Fig. 117.

Die gesättigten Farben Violett und Roth müssen weiter vom Weiss entfernt sein, als ihre weniger gesättigten Complementärfarben, weil nach dem Urtheile des Auges bei der Mischung von Gelbgrün und Violett zu Weiss die Quantität violetten Lichtes viel kleiner ist, als die des gelbgrünen, und wenn das Weiss im Schwerpunkte beider liegen soll, die kleinere Quantität Violett an einem grösseren Hebelarme wirken muss, als die grössere Lichtmenge des Gelbgrün. Uebrigens würden auch hier wieder die Spectralfarben an der Peripherie der Curve, das Purpur auf einer Sehne stehen müssen, Complementärfarben an den entgegengesetzten

> Enden von Sehnen, welche durch den Ort des Weiss gelegt sind, wie bei der kreisförmigen Fig. 114.
> Die Zurueckfuehrung des Farbenmischungsgesetzes auf Schwerpunktconstructionen wurde zuerst von Newton nur als eine Art mathematischen Bildes vorgeschlagen, um die grosse Menge der Thatsachen dadurch auszudrücken, und er stützte sich nur darauf, dass die Folgerungen aus jener Darstellung qualitativ mit den Erfahrungsthatsachen übereinstimmten, ohne dass er quantitative Prüfungen ausgeführt hätte. Dergleichen quantitative Prüfungen sind dagegen in neuester Zeit von Maxwell ausgeführt worden.

In essence, he stated that, if one needs to create colour charts (or circles), in which the areas of the coloured regions correspond to the perceived colours, then the colour space should be like the one shown in Fig. 4.34 (Fig. 117 in the original text of the 1867 edition); in particular, the saturated violet and red colours must be further from white than their less saturated complementary colours, since when mixing yellow, green and violet to get white, the amount of violet light needed is much less. In this colour representation, the spectral colours would have to be on the periphery of the curve, the purple region should be on a chord connecting red and violet, complementary colours on the opposite ends of chords that pass through the location of the white, as in the Newtonian circular representation of Fig. 4.35 (Fig. 114 in the original text of the 1867 edition). In addition, Helmholtz commented that the reduction of the law of colour mixing to constructions of the centre of gravity was first proposed by Newton only as a kind of mathematical representation in order to express the great number of facts by it, and he relied only on the fact that the conclusions from that representation correspond qualitatively with the empirical facts, without being quantitative tested. In contrast, such quantitative tests have recently been carried out by Maxwell. This interesting section of the Handbook also includes a particular insightful paragraph regarding the construction of perceptually consisted colour charts (*Construction der Farbentafel*) in which Helmholtz presents the methodology and the simple mathematics appropriate to create colour charts that correspond to the human perception of colours.

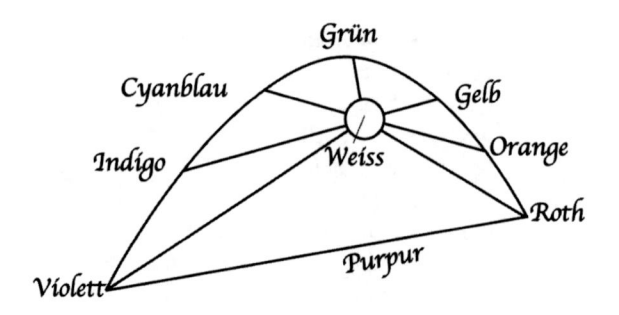

Fig. 4.34 Reproduction of Helmholtz's perceptual colour space representation

Fig. 4.35 Reproduction of a typical Newtonian colour space representation from Helmholtz's 1867 *Handbook*

Last but not least, Helmholtz in his theory emphatically differentiated (using through experiments) the mixing of coloured light to the mixing of pigments, for which he stated that they are two completely different things.

4.16 James Clerk Maxwell

James Clerk Maxwell (1831–1879) was a Scottish scientist that made a significant contribution to mathematical physics with his formulation of the classical theory of electromagnetic radiation that unified the theories of electricity and magnetism, and in essence, founded the field of electrical engineering. He was among those deeply involved in shaping the knowledge about colour, colour sensation and perception. He left a large volume of works regarding optics and colour perception, where he laid out his adoption of the trichromatic model of vision proposed by Young and his quantification and mathematical modelling of phenomena of the optics and colour perception (Maxwell, 1855, 1856a, b, c, 1857a, c, b, 1858, 1860, 1861, 1867, 1869, 1871a, b, 1872, 1874).

In his *Theory of the Perception of Colours* (Maxwell, 1856c), Maxwell defined colour as a function of three independent variables, which he identified as *luminance, hue and tint*. Maxwell declared *tint* as a synonym to *purity*, white being the purest colour, contrary to the definition of saturation by other scientists. Maxwell also recognised that composite colour light consists of infinite variables but adopted Young's theory of trichromacy in that there are three elementary sensations in human vision, by combinations of which all the sensations of colours are produced. He accepted *red, green and violet* as the primary sensations and envisioned all sensations of colours being linear combinations of the primary sensations. Maxwell specifically acknowledged Newton, Young, Helmholtz and Grassmann for their contribution,

> We are indebted to Newton for the original design, to Young for the suggestion of a means of working it out, to Helmholtz for a rigorous examination of the facts on which it rests, and to Professor Grassmann for an admirable theoretical exposition of the subject.

In the subsequent *On the Theory of Compound Colours, and the Relations of the Colours of the Spectrum* (Maxwell, 1860), Maxwell reiterated the suggestion of the curve within the colour triangle following his reasoning based upon Newton, Young and Helmholtz but yet did not provide a quantitative analysis. He unfolded his theory and provided a mathematical analysis and an extensive set of experimental results regarding the observations of compound colours and the accuracy of the observations.

> The investigation of the chromatic relations of the rays of the spectrum must therefore be founded upon observations of the apparent identity of compound colours, as seen by an eye either of the normal or of some abnormal type; and the results to which the investigation leads must be regarded as partaking of a physiological, as well as of a physical character, and as indicating certain laws of sensation, depending on the constitution of the organ of vision, which may be different in different individuals. We have to determine the laws of the composition of colours in general, to reduce the number of standard colours to the smallest possible, to discover, if we can, what they are, and to ascertain the relation which the homogeneous light of different parts of the spectrum bears to the standard colours.

Maxwell suggested that his experiments highlighted a resolution to the dispute about yellow being a primary element of colour, against this hypothesis, since his experiments showed that the composition of yellow from red and green is indistinguishable to pure yellow by the observers, and only a prism could expose the difference.

Almost a decade later, Maxwell in his treatise *On Colour Vision* Maxwell (1871a, 1872) summarised the current knowledge (basically his recognition of Newton's and Young's work) and his contribution to the domain. He expressed his alignment with the theory that humans are capable of three different colour sensations, which light of any kind excites in different proportions in order to produce all the varieties of sensed colours. To Maxwell *seemed almost a truism to say that colour is a sensation.* To him *colour is related to human physiology and by no means to the nature of light.* He opens this paper by stating that

> All vision is colour vision, for it is only by observing differences of colour that we distinguish the forms of objects. I include differences of brightness or shade among differences of colour.

He escalates his reasoning by emphatically stating that

> The science of colour must therefore be regarded as essentially a *mental science*. It differs from the greater part of what is called mental science in the large use which it makes of the physical sciences, and in particular of optics and anatomy. But it gives evidence that it is a mental science by the numerous illustrations which it furnishes of various operations of the mind.

Regarding the apparent colour of objects, Maxwell stated that when objects are illuminated by white light, they separate that light into its components absorbing some and scattering others. In addition, using the analogy of the sensation of sound, in which humans are able to recognise the components of a composite sound (analysed into elementary sensations), Maxwell explained that this is not the case with colour, the sensation of (composite) colour being a *single thing* and cannot be decomposed into its elementary sensations (or the sensation of its elementary components). Further, Maxwell affirmed that it is easy to experimentally prove that *the quality of colour can vary in three and only three independent ways*, restating the three independent variables being *hue, tint and shade* (in a slightly different way, as he initially proposed the terms luminance, hue and tint), and emphatically highlighting that if one adjusts one colour to another, so as to agree in hue, tint and shade, the two colours would be absolutely indistinguishable.

Maxwell was not fully satisfied with the experimental results derived from colour wheels so he devised a series of instruments, the 'light boxes' (Longair, 2008) or 'colour boxes' (MacAdam, 1970) to be able to make accurate measurements on colour perception. A schematic diagram and the principle of operation of Maxwell's light box is shown in Fig. 4.36. Reference white light is shone onto the top slit as well as three adjustable lights (blue, green and red) at corresponding slits at $\mathcal{B}$. The reference white is guided through the box to the eyes of the observer, through a set of mirrors C. The three adjustable lights are being mixed and also guided through mirrors and prisms to the observer. There, according to the width of the slits at $\mathcal{B}$ the tested composite light can be subjectively compared to the reference light. While the observer reports a difference, the test lights are being adjusted by means of the widths of the corresponding slits. When the observer reports a perfect match, the position of the slits at $\mathcal{A}$ and the width of the slits at $\mathcal{B}$ are used to register a *colour equation*. Maxwell carried out a number of such experiments, which resulted in accurate information about the composition of various colours and provided the first types of modern chromaticity diagrams, which define how different colours can be synthesised from chosen primary colours.

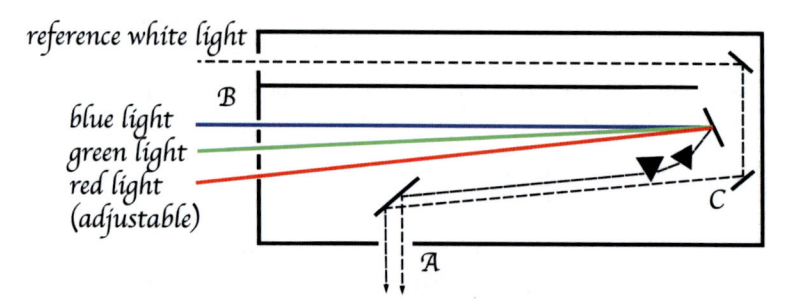

Fig. 4.36 A schematic diagram and the principle of operation of Maxwell's light or colour box

Fig. 4.37 The famous tartan ribbon, the first ever permanent coloured photograph captured by using Maxwell's technique in 1861

It should be noted that in 1861 Maxwell was able to capture *the first permanent coloured photograph* (Fig. 4.37). Maxwell's technique involved the use of three positive photographic plates and three filters (red, green and blue-violet). The three positives that were produced, were projected through the same filters onto a screen and thus combined into a reasonably fully coloured image.

4.17 Ewald Hering

Karl Ewald Konstantin Hering (1834–1918) was born in Alt-Gersdorf, Kingdom of Saxony. He was a German physiologist with a main interest in the physiology of the visual system and colour vision. He proposed what is known as the *opponent colour theory*, which was a significant paradigm shift in comparison to the standard trichromatism accepted by most scientists before him. His work, among other subjects, included studies of binocular vision, hyper-acuity and eye movement. In

1879 he made a significant proposition regarding the visual direction that was later named after him (*Hering's law* or *law of visual direction*), which described the perceived visual direction of a natural object in relation to the observer, having in mind that there are two visual sensors (binocular vision). According to this law, *everything that is in the line of sight of each of the eyes appears mixed in one and only one virtual egocentric direction*. This is like having a single eye, referred to as the *cyclopean eye*, positioned in the middle of the two eyes. Hering published important work on the spatial sensing and the movement of the eyes (E. Hering, 1879), on the binocular vision (E. Hering, 1868), on visual acuity (K. E. K. Hering, 1899; Strasburger et al., 2018) and colour theory (K. E. K. Hering, 1964, 1878, 1920).

In his 1892 *Grundzüge der Lehre vom Lichtsinn* (*Outlines of a Theory of the Light Sense*) (K. E. K. Hering, 1964, 1878, 1920), Hering dismissed the prevailing theory developed by Young-Maxwell-Helmholtz, by which colour vision is based on a model of three-dimensional colour sensing (three primary colours and three types of photoreceptors). He dismissed this theory, primarily based on experiments in colour adaptation and observations in the usage of colour-related linguistic terms; the latter is a very interesting argument, as one may easily accept that there cannot be a whitish-black, or a reddish-green or even a yellowish-blue colour, as it is natural to accept, for example, a yellowish-red, or a greenish-blue, or even a greenish-yellow. Since there are no linguistic terms to define some composite colours, while there is a multitude of such terms for all other colours, this is strong evidence that those particular colours cannot exist. Hence, Hering concluded that colour vision must be based on detection of colour opponency, and particularly the one of red-green, yellow-blue and white-black. This way, he indirectly defined a four-colour system, by proposing that since yellow is not perceived as a red-green mixture (although it can be produced by such a mixture), it should be regarded as one of the colours to define the new opponent model. He found this mechanism to be more efficient, also because it was already suggested by models and evidence that the various receptors in the eye exhibit an overlapping sensitivity (see for example Fig. 4.33).

Hering's *Grundzüge der Lehre vom Lichtsinn* begins with a statement that emphasises the role of colour perception, highlighting that our world of vision consists only of different colours, and the things we see are nothing other than colours of different types and shapes.

Unsere Sehwelt besteht lediglich aus verschieden gestalteten Farben, und die Dinge, so wie wir sie sehen, d.h. die Sehdinge, sind nichts anderes als Farben verchiedener Art und Form.

He also emphasised the role of perception over sensing, by reminding that the whole world of vision and its content is a creature of the *inner eye*, a name for the complete nervous organ of vision (retina, optic nerve and the related parts of the brain), in contrast to the dioptric apparatus, the *outer eye*. The creative capacity of the inner eye creates these colour structures under the compulsion of the stimuli that receives from the radiation sent into the eye by real external objects.

> Die ganze Sehwelt mit ihrem Inhalt ist ein Geschöpf unseres inneren Auges, wie wir das nervöse Sehorgan (Netzhaut, Sehnerv und die bezüglichen Hirnteile) nennen können, im Gegensatze zu dem dioptrischen Apparat als dem äußeren Auge. Das schöpferische Vermögen unseres inneren Auges schafft jene Farbengebilde unter dem Zwange der Anregungen, welche es durch die von den wirklichen Außendingen in unser Auge geschickten Strahlungen erhält.

Based on this, he clarified that the real world and the world accessible by the senses should be regarded as totally independent. He also made clear that perception of colour is bound by the process of adaptation, thus an object may appear to have a different colour in different viewing conditions. In addition, he noted that colours appear different when perceived by different parts of the retina, and particularly in central and peripheral vision.

Reproductions of Hering's colour circles are shown (a) in Fig. 4.38, in which the four primary colours are shown, along with their interaction and contribution to the production of all other colours, and (b) in Fig. 4.39, in which some colours are drawn in a classic colour circle corresponding to Hering's model, where no colour is clearly reddish and greenish, nor yellowish and bluish at the same time, red and green are just as mutually exclusive as yellow and blue.

> Keine Farbe ist deutlicherweise rötlich und grünlich, keine gelblich und bläulich zugleich, Röte und Grüne schließen sich ebenso aus wie Gilbe und Bläue.

Regarding the white-black opponent pair, he suggested that although this pair has a natural gradation from one to the other through the scale of grey, there is no such a gradation for the other opponent pairs, yellow-blue and red-green, which need to fade completely to grey in order to go from one to the other. It is there that Hering defined the opponent colours (*die Gegenfarben*), as the mutually exclusive pairs that cannot, in any "normal" circumstances, appear at the same time.

> Da also Röte und Grüne, bezw. Gilbe und Bläue in keiner Farbe gleichzeitig deutlich sind, sich vielmehr gegenseitig auszuschließen scheinen, habe ich dieselben als Gegenfarben bezeichnet. Hiermit soll zunächst lediglich die Art ihres Vorkommens gekennzeichnet sein ohne jede Beziehung auf irgendwelche Erklärung.

In his colour circle (Fig. 4.39), any pair of diametrically opposite colours (yellowishred and bluish-green) should be considered doubly opponent, in contrast to any

Fig. 4.38 Reproduction of
Hering's colour circle with
opponent colours

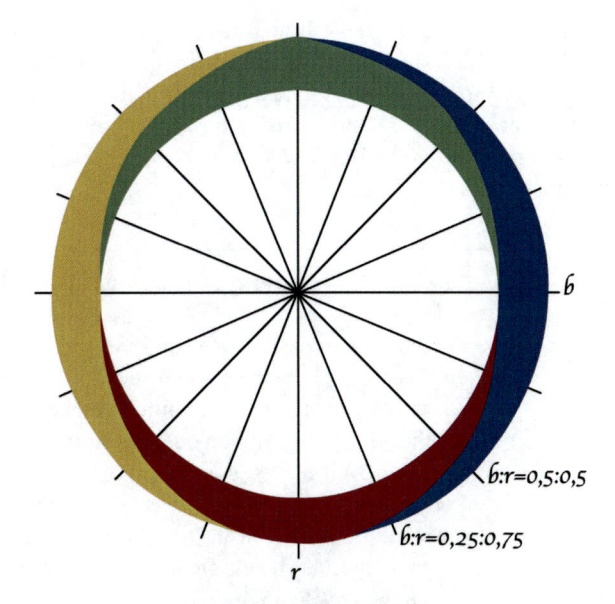

Fig. 4.39 Reproduction of
Hering's colour circle

colour pairs symmetrical around a primary colour (reddish-yellow and greenish-yellow), which should be considered singly opponent.

Hering found it most interesting that there are those particular pairs of colours (the opponents), for which there cannot be any intermediate mixtures and concluded this should be hardwired in visual perception, leading to his opponent colour perception theory.

> Es erscheint von vornherein höchst auffällig, dass es z.B. zwischen Rot und Grün nicht ebenso eine Reihe bunter Zwischenfarben giebt, wie zwischen Rot und Gelb oder zwischen Rot und Blau, dass es also keine Farben giebt, welche uns in ähnlicher Weise zugleich rötlich und grünlich erscheinen, wie das Orange zugleich rötlich und gelblich oder das Grau zugleich weißlich und schwärzlich. Wir dürfen daraus schließen, dass im inneren Auge ein physiologischer Prozess, dessen psychisches Korrelat von gleichzeitig deutlicher Röte und Grüne bezw. Gilbe und Bläue wäre, entweder überhaupt nicht oder nur unter ganz besonderen, ungewöhnlichen Bedingungen möglich ist.

Hering was also interested in the non-linear response of vision and he tried to gather data to support a model for this phenomenon. Figure 4.40 shows a reconstruction of his plot of perceived brightness against luminance, in which the two curves correspond to a medium grey at $1/2$ and $1/3$ of the $W - S$ distance (black-to-white normalised perceived brightness scale). Clearly, the perceived brightness increases rapidly with little luminance increase at low luminance conditions, and then slowly and asymptotically converges to maximum brightness.

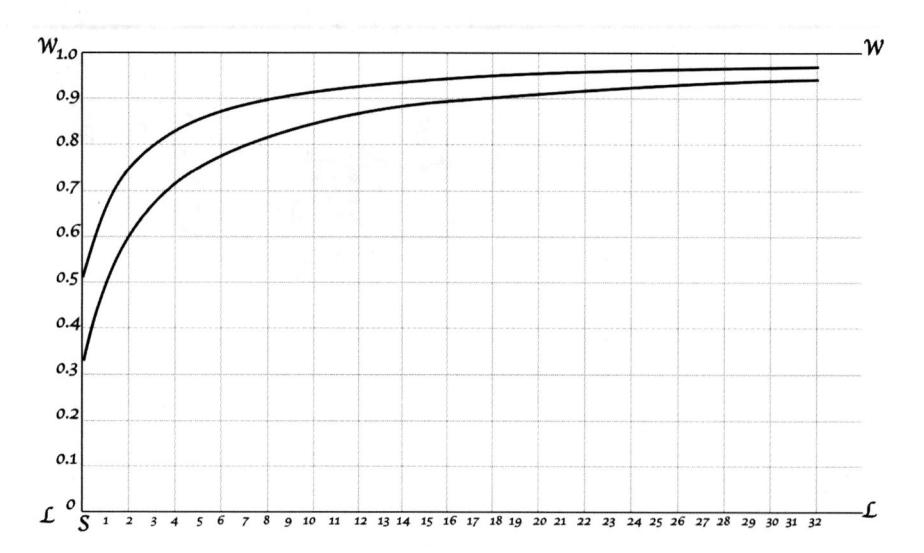

Fig. 4.40 Reproduction of Hering's plot of perceived brightness against luminance

4.18 Gottlob Frege

Friedrich Ludwig Gottlob Frege (1848–1925) was born in Wismar, Grand Duchy of Mecklenburg-Schwerin, German Confederation. He was a philosopher, logician and mathematician. He is considered to be the father of analytic philosophy in language, logic and mathematics. Interestingly, he was not so famous in his lifetime and his work became known later through other eminent thinkers, like Bertrand Russell (1872–1970) and Ludwig Wittgenstein (1889–1951), and is considered to be among the greatest mathematical philosophers and logicians, on the scale of Aristotle. He left very important writings, and although not directly connected to the topic of this treatise, he still deserves to be included due to the indirect implications of his theories.

His most famous works include the 1879 *Begriffsschrift, Eine der Arithmetischen Nachgebildete Formelsprache des Reinen Denkens* (*Begriffsschrift, a Formula Language, Modeled upon that of Arithmetic, for Pure Thought*) (Frege, 1879), the 1884 *Die Grundlagen der Arithmetik* (*The basics of arithmetic*) (Frege, 1884), the 1892 *Über Sinn und Bedeutung* (*On Sense and Reference*) (Frege, 1892), the 1893 *Grundgesetze der Arithmetik* (*Basic laws of arithmetic*) (Frege, 1893) and the most influential 1918 *Der Gedanke: Eine logische Untersuchung* (*The thought: a logical investigation*) (Frege, 1918). Frege, in his attempt to turn mathematics into an application of logic, he invented a new symbolic language to describe mathematics based on pure logic, which he named *Begriffsschrift* (conceptual notation, or ideography). It was like inventing a programming language in more modern terms.

Frege's work that is most relevant to a theory of visual perception is the 1892 *Über Sinn und Bedeutung* (*On Sense and Reference*) (Frege, 1892; McCarty et al., 2000). In this treatise, Frege made extensive use of linguistics and philosophy to present the basic threefold nature of any object in human perception (Fig. 4.41). His theory distinguishes an object (reference) to its name (sign) and its cognitive content (sense), with extensions from simple names to sentences and the notion of the truth. Practically, in his theory, Frege accepts the objectivity of an entity by means of

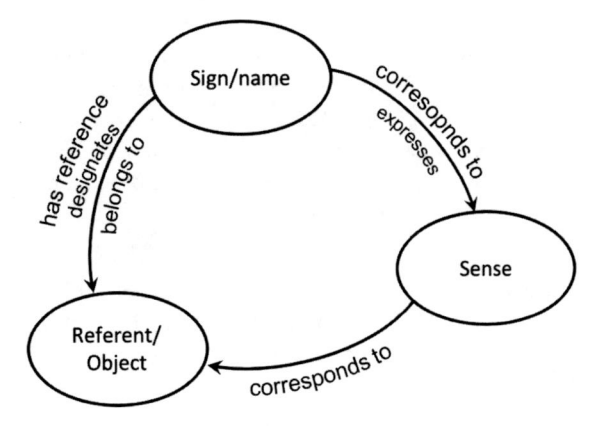

Fig. 4.41 Frege's famous Sign-Sense-Reference

commonly shared subjective senses, like considering something to be objectively true if the senses of multiple observers agree. On the other hand, the sense is a purely subjective experience and only due to strife for the truth, sense reaches out to find a reference.

Frege clarified even further his philosophical views in his 1918 *Der Gedanke: Eine logische Untersuchung* (*The thought: a logical investigation*) (Frege, 1918). There he differentiated senses from objects and ideas, somehow reviving the Platonic world of ideas (ideals). He discussed the paradox of the meaning of a colour name when comparing the sensation of colour between a person with normal vision and a colour-blind. To the colour-blind person, the green strawberry plant leaves and the red strawberries will appear in the exact same colour. Then, Frege wonders what the name of that colour might be when the normal vision person already names the leaves green and the strawberries red.

> Mein Begleiter und ich sind überzeugt, daß wir beide dieselbe Wiese sehen; aber jeder von uns hat einen besonderen Sinneseindruck des Grünen. Ich erblicke eine Erdbeere zwischen den grünen Erdbeerblättern. Mein Begleiter findet sie nicht; er ist farbenblind. Der Farbeneindruck, den er von der Erdbeere erhält, unterscheidet sich nicht merklich von dem, den er von dem Blatt erhält. Sieht nun mein Begleiter das grüne Blatt rot, oder sieht er die rote Beere grün? oder sieht er beide in einer Farbe, die ich gar nicht kenne? Das sind unbeantwortbare, ja eigentlich unsinnige Fragen.

For Frege, thought is the glueing factor between the world and its perception. The reception of sense impressions is not sufficient for seeing things. He makes clear that in the perception of a tree, physical, chemical, and physiological processes slip between the tree and the imagination (or thought). However, only processes in the nervous system are directly related to consciousness; and every observer of the tree has an individually special process in a special (or unique) nervous system.

> Zwischen den Baum und meine Vorstellung schieben sich physikalische, chemische, physiologische Vorgänge ein. Mit meinem Bewußtsein unmittelbar zusammen hängen aber, wie es scheint, nur Vorgänge in meinem Nervensystem; und jeder Beschauer des Baumes hat seine besonderen Vorgänge in seinem besonderen Nervensystem.

Most interestingly, Frege concluded that any observer may only be conscious of the end process of perception and cannot make objective connections of perceived impressions with independent stimuli that produce the sensory impressions in the first place.

> Wir glauben, daß ein von uns unabhängiges Ding einen Nerv reize
> und dadurch einen Sinneseindruck bewirke; aber genau genom-
> men, erleben wir nur das Ende dieses Vorganges, das in unser
> Bewußtsein hereinragt.

4.19 Johannes von Kries

Johannes Adolf von Kries (1853–1928) was born in Freiburg, Germany, He was a physiological psychologist, particularly interested in the neural mechanisms of perception. He was the founder of the *duplicity theory* of vision, in which two types of photoreceptors are responsible for vision; these are the rods (or rod cells) that are efficient in low light conditions and the cones, which are three types of cells efficient at brighter lighting conditions. Apart from his main interest in the mechanism of visual perception, he made notable contributions to the foundations of the theory of probability. Von Kries is considered to be Helmholtz's warmest advocate (Turner, 1994).

Von Kries' relevant contribution can be found in two publications, (a) the 1878 *Beitrag zur Physiologie der Gesichtsempfindungen* (*Contribution to the physiology of visual sensations*), included in Volume I & II of the *Archiv für Anatomie und Physiologie*, edited by Wilhelm His, Wilhelm Braune and Emil Du Bois-Reymond (Kries, 1878) and (b) the 1905 *Die Gesichtsempfindungen* (*The visual sensations*), included in Volume III (Physiology of the Senses) of the *Handbuch der Physiologie des Menschen*, edited by Wilibald Nagel (von Kries, 1905).

In Kries (1878), he stated that one of the most important tasks for a theory of visual sensations is to explain why the diversity of our sensations is much smaller than that of light stimuli. While the latter can be varied in infinite abundance, the variety of visual sensations is only threefold. This can be formulated as follows: if any visual sensation can be produced by the action of α, β, γ quantities of the three primary types of light A, B, C, then every possible continuous change of the sensation can be produced by the constant change in the quantities α, β, γ.

> Für die Theorie der Gesichtsempfindungen ist es eine der wichtig-
> sten Aufgaben, zu erklären, warum die Mannichfaltigkeit unserer
> Empfindungen eine viel geringere ist als die der Lichtreize.
> Während die letzteren in unendlicher Fülle variirt werden kön-
> nen, ist die Mannichfaltigkeit der Gesichtsempfindungen, wie
> man zu sagen pflegt, eine nur dreifach ausgedehnte. Den Satz,
> auf welchen es hier ankommt, können wir so formuliren: Wenn
> eine beliebige Gesichtsempfindung hervorgebracht werden kann
> durch die Einwirkung einer Mischung der Quantitäten α, β, γ, der

> drei einfachen Lichtarten A,B,C, so kann jede überhaupt mögliche continuirliche Aenderung der Empfindung hervorgebracht werden durch die stetige Aenderung der Quantitäten α, β, γ.

Von Kries affirmed that the two most important theories of visual sensation during that period, advanced by Thomas Young, renewed by Helmholtz and Maxwell, and those proposed by Hering, all agreed that sensation is determined by the values of three independent variables, which are supposed to consist in the various intensities of a limited number of simple nervous processes.

> Die beiden gegenwärtig bedeutendsten Theorien der Gesichtsempfindungen, die von Thomas Young aufgestellte, (von Helmholtz und Maxwell erneuerte), und die kürzlich von E. Hering gegebenemachen zur Erklärung dieser Thatsache übereinstimmend die Annahme, dass die Empfindung bestimmt sei durch die Werthe dreier unabhängig veränderlicher Functionen, welche bestehen sollen in den verschiedenen Intensitäten einer beschränkten Anzahl einfacher nervöser Vorgänge. Wenn auf jede dieser drei Functionen drei verschiedene Lichter einen verschiedenenEinfluss üben, so erklärt sich daraus unmittelbar der obige Satz.

In von Kries (1905), von Kries provided a concise theory of visual sensation. He began his treatise by providing background knowledge on the subject and by analysing the laws of colour mixing, with references to Newton and Grassmann. He also provided a representation of the colour chart of his time (Fig. 4.42), as the already established curved diagram of the spectral colours connected with the horizontal line of purples.

Given a mixture of, say greenish-blue (Gbl) and red (R) one may generate the same sensation with a mixture of green (Gr) and violet (V), which may result in an interesting mathematical formulation.

Fig. 4.42 Reproduction of Von Kries' colour chart

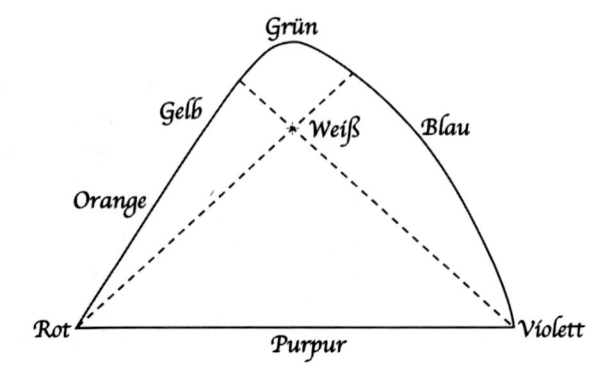

$$\alpha Gbl + \beta R = \gamma Gr + \delta V \Rightarrow$$
$$\alpha Gbl = \gamma Gr + \delta V - \beta R \tag{4.10}$$

By using the second formulation, it is shown that, provided one accepts negative values, all sensations can be produced by mixtures of three lights.

Von Kries moved on to the study of complementary colours, beginning with mixtures of red and green. He adopted the definition by which complementary should be called the colours that when mixed in certain proportions result in colourless (white) sensation, and adopted and reiterated Helmholtz's table of complementary colours and graphical representations. Furthermore, von Kries provided his own table of complementary colours. On this, he briefly reviewed the Young-Helmholtz trichromacy theory, which he accepted. Nevertheless, he also confirmed the validity of Hering's opponent colour theory, which he described and commented, regarding it as an embodiment of a four-colour theory.

> Es ist, um Mißverständnisse zu vermeiden, wichtig, sie auseinander zu halten von der weit allgemeineren, oben als Vierfarbentheorie bezeichneten Anschauung, welche letztere, wenn sie mit dem Namen eines bestimmten Autors in Verbindung gebracht werden soll, wohl am ehesten an den Auberts zu knüpfen wäre. Die Theorie Herings ist eine auf gewisse allgemein biologische Vorstellungen gestützte Ausgestaltung der Vierfarbentheorie.

Von Kries was largely interested in the light adaptation characteristics of vision. He made a distinction between day-time vision (*Tagessehen*) and night-time vision (which he called the twilight vision–*Dämmerungssehens*) and studied the limits of those types of vision. He experimented with colour-blind people to identify the sensitivity of the eyes and produced light sensitivity curves for night-time vision, a reproduction of which is shown in Fig. 4.43; the graph shows the perceived brightness for various wavelengths of light for the case of night-time vision of a normal observer (solid line) and for the case of a colour-blind (dashed line), under the illumination of gaslight. This observation led him to support the dual sensor vision (cones and rods) and also the duplicity theory of vision. Since colour-blind people can perceive light of various wavelengths (which is the decisive feature for colour perception), the rods, which are active during night-time vision (he called them the twilight organs–*Dämmerungsorgane*), should not provide any information regarding the colour representation of light. This, he added, was also supported by other experiments regarding the visual acuity and response time to fast-changing stimuli. Von Kries broaden his study by examining the impact of the position in the field of view on the perception of colour (linked with the position of the cones and rods in the retina) and concluded that, as objects move away from the centre of the field (in any direction), colour is lost to a point that all objects appear in shades of grey in the extreme periphery. Another interesting graph reproduction of von Kries' original sensitivity curves is shown in Fig. 4.44, in which he showed the brightness percep-

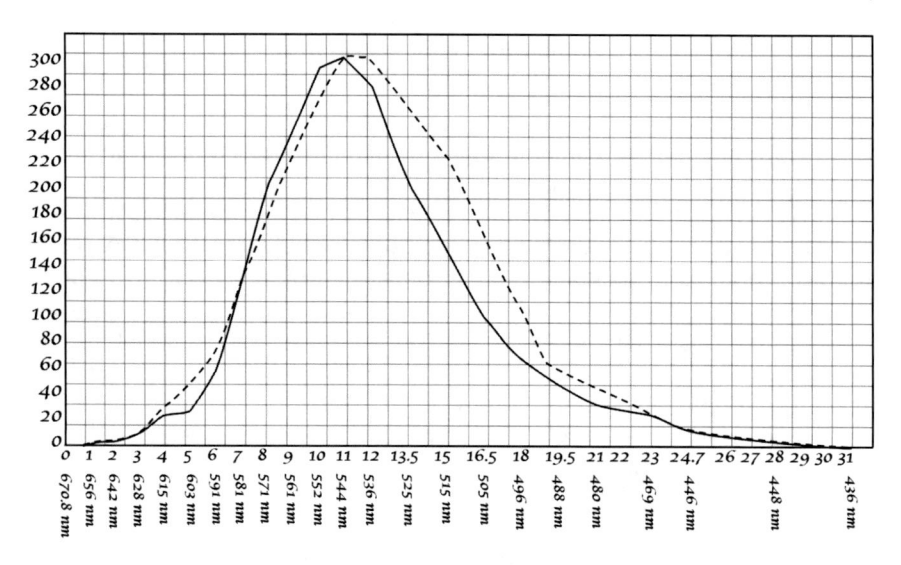

Fig. 4.43 Reproduction of Von Kries' light sensitivity curves

Fig. 4.44 Reproduction of
Von Kries' light sensitivity
curves

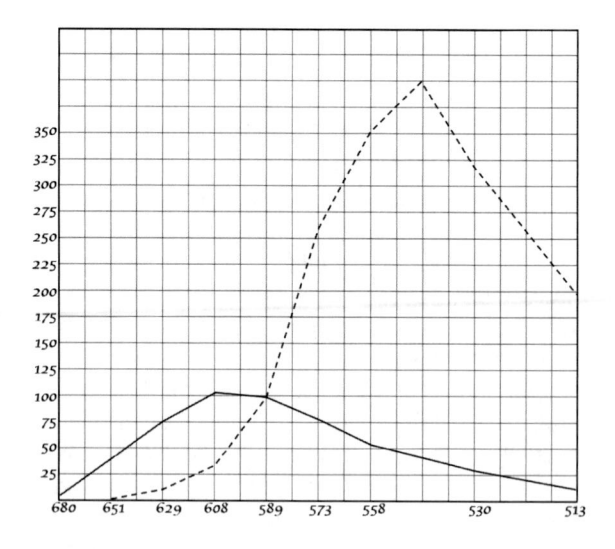

tion against the wavelength for either the light-adapted peripheral vision (solid line) or the normal night-time vision (dashed line).

One of von Kries' most important contributions is in the field of *adaptation*. As the term suggests, adaptation is the ability of the retina to adjust its sensitivity to be able to function efficiently in various lighting conditions (practically, by redefining the bases of what should be considered to be 'black'). In accordance with the duplicity theory that von Kries supported, adaptation takes place as a sensitivity transition from rods to cones (and vice versa) and within each of the two photoreceptors to around nine (9) orders of magnitude. What von Kries suggested in *Der Koeffizientensatz* (the set of coefficients) (von Kries, 1905) was that the process of adaptation is somewhat linear and depends on a coefficient per photoreceptor (with values in [0, 1]).

> Es liegt nämlich nahe, anzunehmen, daß, soweit die Wirkung äußerer Reize in Frage kommt, die Stimmung, sei es des Sehorgans in toto, sei es einzelner Bestandteile, sich als eine größere oder geringere Erregbarkeit gegenüber jenen Reizen geltend machen wird, und zwar so, daß der Erfolg sich immer etwa nach einem Produkt α R richtet, wo R den Reizwert, α aber die für diesen Erfolg bestehende Disposition oder die für diese Reizart vorhandene Erregbarkeit bezeichnen würde.

Practically, this means that for three photoreceptors, sensitive in red (R), green (G) and blue (B) light (R, G, B being the stimuli), there are three coefficients α, β, γ, which regulate the response of the retina. If R_r, G_r, and B_r are sensitivities under a reference illuminant then this relationship is expressed in (4.11).

$$R_r = \alpha R$$
$$G_r = \beta G \qquad \qquad (4.11)$$
$$B_r = \gamma B$$

The coefficients α, β, and γ are the *von Kries coefficients* and correspond to the reduction in sensitivity of the three-cone mechanisms due to chromatic adaptation (relative to reference sensitivity) and are constant for all pairs of corresponding colours. This, in effect, means that any colour sensation will remain the same under various lighting conditions and is another expression of what is known as *colour constancy*. Von Kries expressed this linear relationship by stating another typical linearity criterion.

> Es müßte nämlich dann, wenn L_1 auf einer Netzhautstelle den gleichen Erfolg auslöst wie L_2, an einer anderen, und ebenso M_1, auf die erstere wirkend, den gleichen Effekt wie M_2 an der anderen, jedesmal auch $L_1 + M_1$ hier die gleiche Wirkung haben müssen wie $L_2 + M_2$ dort.

If L_1 triggers the same effect on one site of the retina as L_2 on another, and also M_1, acting on the former, it would have to have the same effect as M_2 on the latter, then each time $L_1 + M_1$ must also have the same effect on the first site as $L_2 + M_2$ on the second. This is a rather significant statement and this is what von Kries called the *Coefficient Theorem (Koeffizientensatz bezeichnen)*. Recognising the approximating nature of this theory, von Kries mentioned that this is an approximate theory and cannot be verified in all experiments. Nevertheless, it has been extensively used in practical applications and particularly in digital camera technology for the process of estimating the white balance.

4.20 Arthur König

Arthur Peter König (1856–1901) was born in Krefeld, a city in North Rhine-Westphalia, Germany. His life work was on physiological optics. He studied under Hermann von Helmholtz, and eventually became his assistant. His work on optics was rather fruitful with published papers of significant importance, among which the works co-authored with Conrad Dieterici *Die Grundempfindungen und ihre Intensitäts-Vertheilung im Spectrum* (*The fundamental sensations and their sensitivity distribution in the Spectrum*) (König & Dieterici, 1886) and *Die Grundempfindungen in normalen und anomalen Farbensystemen und ihre Intensitätsverteilung im Spektrum* (*The fundamental sensations in normal and abnormal colour systems and their sensitivity in the spectrum*) (König & Dieterici, 1892). This work is of particular interest since it is a foundation work on the sensitivity of the human rod and cone visual system, which improved the earlier measurements by Maxwell, using more advanced equipment and procedures. Moreover, this work made it possible to propose a theory about defects in vision (dichromacy and colour blindness) and their connection with the absence of cone types in the retina. The data resulted by König's work were replaced some 30–40 years later by the more accurate data by John Guild and William David Wright, which eventually became the foundation of the modern CIE colour system. König, apart from his own work, he left a significant editorial work among which the *Beiträge zur Psychology und Physiologie der Sinnesorgane* after Helmholtz.

König and Dieterici (1892) made extensive measurements for the determination of the qualitative and quantitative nature of *white light*, for example with high-resolution relative estimates of the sunlight in relation to gaslight in a wide range of wavelengths. In addition, they performed estimates of the dispersion and interference spectrum of the gaslight and provided detailed illustrations of the visual sensation in monochromatic and dichromatic colour systems. They called their sensitivity curves the *elementary sensation curves* produced by several researchers. König and Dieterici made detailed measurements of the complementary colours under sunlight and gaslight and a number of experiments on the determination of the visual sensitivity by means of the complementary colours. They were among the very first to provide trustworthy estimates of the colour sensitivity curves, which are shown in Fig. 4.45; the graph presents their elementary RGV (V for violet) sensation curves, which were created by using their measurements (in Tables XVI and XVII) for the case of sunlight. The measurements were interpolated (typical spline interpolation) for a smoother presentation. The bold curves correspond to König's measurements, whereas the lighter curves correspond to Dieterici's measurements, for normal trichromatic vision.

According to their theory, basic sensations are a product of elementary sensations in a linear manner. Thus, the basic sensations $\Re$, $\mathfrak{G}$, $\mathfrak{B}$ for a normal trichromatic colour system can be defined (in their original notation) as

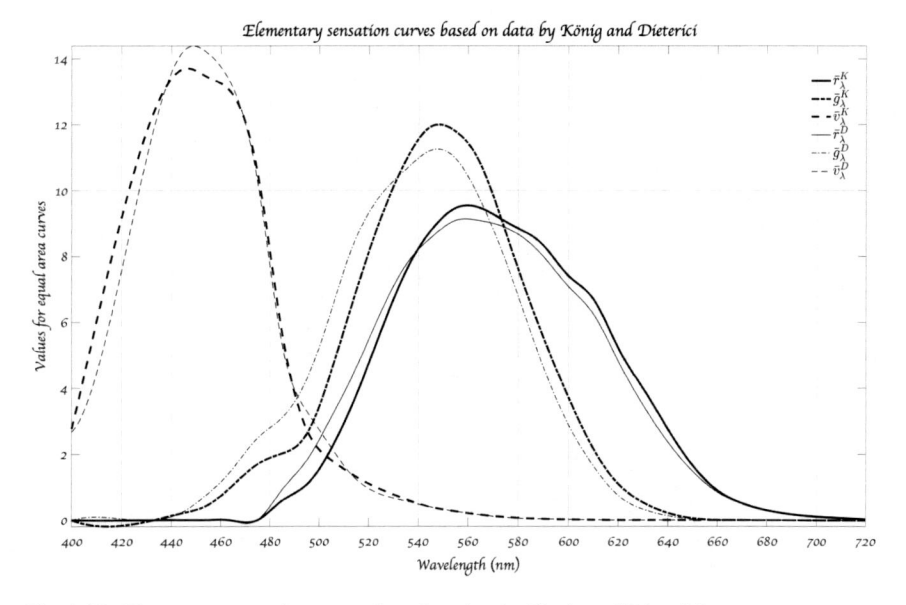

Fig. 4.45 Elementary sensation curves based on data by König and Dieterici

$$\mathfrak{R} = a' \cdot R + b' \cdot G + c' \cdot V$$
$$\mathfrak{G} = a'' \cdot R + b'' \cdot G + c'' \cdot V \qquad (4.12)$$
$$\mathfrak{B} = a''' \cdot R + b''' \cdot G + c''' \cdot V$$

They also proposed the normalised form

$$\mathfrak{R} = \frac{a' \cdot R + b' \cdot G + c' \cdot V}{a' + b' + c'}$$
$$\mathfrak{G} = \frac{a'' \cdot R + b'' \cdot G + c'' \cdot V}{a'' + b'' + c''} \qquad (4.13)$$
$$\mathfrak{B} = \frac{a''' \cdot R + b''' \cdot G + c''' \cdot V}{a''' + b''' + c'''}$$

again, for normal trichromacy, and proposed the corresponding coefficient values as follows

$$a' = 1 \qquad b' = -0.15 \; c' = 0.1$$
$$a'' = 0.25 \; b'' = 1 \qquad c'' = 0 \qquad (4.14)$$
$$a''' = 0 \qquad b''' = 0 \qquad c''' = 1$$

They went further and provided a definition of the standard observer (normal trichromat) from their data. Figure 4.46 shows the normal observer curves that can be created by interpolation of their sparse data. In addition, they were able to use

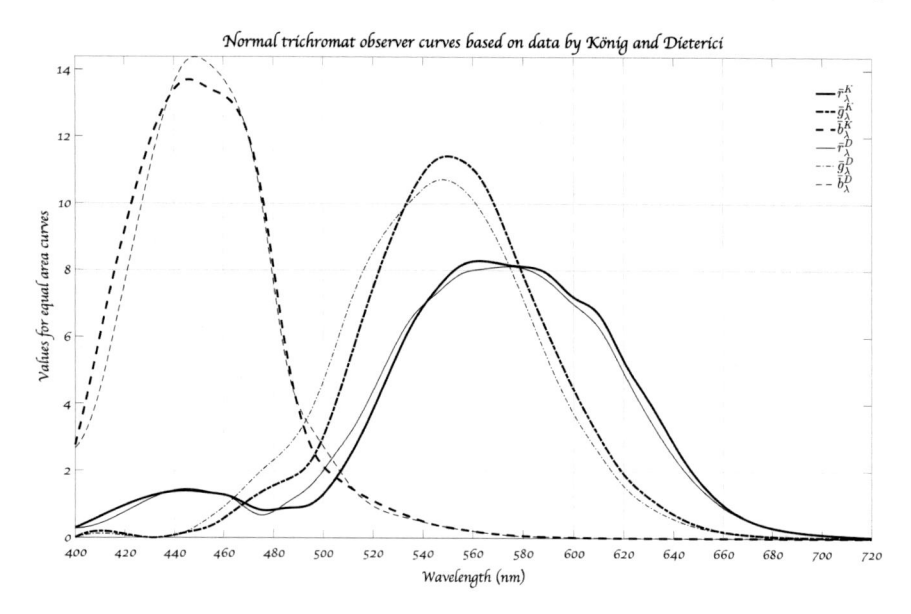

Fig. 4.46 Normal trichromat observer curves based on data by König and Dieterici

Fig. 4.47 König and
Dieterici colour space in the
classic colour triangle

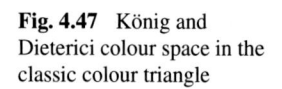

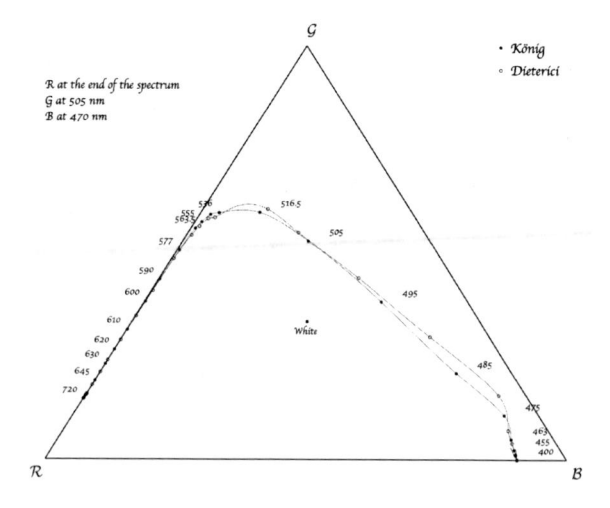

those data and place the position of the various wavelengths of monochromatic light
into the typical R-G-B colour triangle, as shown in Fig. 4.47.

Last but not least, *they confronted Hering's opponent theory* by stating that the
optic nerves should somehow resemble the motor nerves in that they either rest or
become excited and do not exhibit antagonistic behaviour.

4.21 Ramón y Cajal

Santiago Ramón y Cajal (1852–1934) was born in Petilla de Aragón, Navarre, Spain. He was a neuroscientist, pathologist, and histologist specialising in neuroanatomy and the central nervous system, with original investigations of the microscopic structure of the brain that made him a pioneer (actually the father) of neuroscience.

It was the year 1887 that Ramón y Cajal was awarded a professorship in Barcelona, where he got accustomed to Golgi's staining method that made it possible to visualise neural tissue.[31] He worked and even improved this method, as it became central to his work. This unlocked the potential to investigate the elusive (at that period) structure of the central nervous system. The result was an extensive mass of detailed hand-made drawings of neural tissue that depicted the *arborisations* of neural cells, covering many species and parts of neural systems.

Since 1880 he had been publishing important scientific works, among which his 1899 *Textura del Sistema Nervioso del Hombre y de los Vertebrados* (*Textbook on the nervous system of man and the vertebrates*) (Ramón y Cajal, 1899) is considered to be his Opus Magnus. Ramón y Cajal realised that to unravel the mysteries of the nervous system, and particularly the brain, is to always look at the big picture, the system as a whole. In addition, he supported the theory that a fundamental characteristic of any brain is the presence of organised neural circuitry (Llinás, 2003). His trend towards the integration into a global framework is particularly evident in the *Textura del Sistema Nervioso del Hombre y de los Vertebrados*. Among the most important proposals was Cajal's suggestion that the organisation of the central nervous system was the result of natural optimisation, by which the system saves space, time and material (*el ahorro del espacio, el ahorro del material, el ahorro del tiempo*). A compact presentation of his work can be found in his 1894 *The Croonian lecture-La fine structure des centres nerveux* (Ramón y Cajal, 1894a), whereas other important publications include his 1894 French edition *Les nouvelles idées sur la structure du système nerveux chez l'homme et chez les vertébrés* (*The new ideas on the structure of the nervous system in humans and vertebrates*) (Ramón y Cajal, 1894b) and the 1909 and 1911, also French editions of the *Histologie du système nerveux de l'homme & des vertébrés*, Tome I & II (*Histology of the nervous system of man & vertebrates*, Volumes I & II) (Ramón y Cajal, 1909, 1911). Cajal published a large volume of scientific articles on the fine structure of the nervous system in both French and Spanish. Of particular interest for the purposes of this treatise is his work towards the understanding of the structure and function of the nervous system that is relating to vision.

---

[31] At that time, there were no staining techniques usable for the study of the nervous tissue. Staining techniques use special substances to cause the increase of the contrast among various tissue types, largely useful for optical inspection under the microscope. Golgi, around 1873, discovered a method suitable for staining nervous tissue in black, which he called *la reazione nera* (black reaction), which is typically called Golgi's method or Golgi's staining. This was a major breakthrough in neuroscience that made neural tissue visualisation possible under the microscope.

In the 1894 *Croonian Lecture* (Ramón y Cajal, 1894a), Cajal provided a detailed diagram of the retina of the human eye and a very brief and compact definition of the retina,

> On peut, malgré sa complication, considerer la rétine comme un ganglion nerveux formé par trois rangées de neurones ou de corpus–cules nerveux; la première rangée renferme les cônes et les bâtonnets avec leurs prolongements descendants formant la couche des grains externes; la seconde est constituée par les cel-lules bipolaires, et la troisième est due à la réunion des corpus-cules ganglionnaires. Ces trois séries d'éléments s'articulent au niveau des couches dites moléculaires ou réticulaires et internes.

He tried to simplify the description of the complicated structure by stating that the retina can be considered as a nervous ganglion formed by three rows of neurons or nervous corpuscles, including the cones and the rods with their descending extensions forming the layer of the outer grains, the layer of the bipolar cells, and the layer of the union of the ganglionic corpuscles. Figure 4.48 shows a graphical representation of a part of the human retina, based on the original drawing published as Fig. 4 in Ramón y Cajal (1894a). In this representation, according to Cajal's description, *A* denotes the cones from the region of the fovea centralis, *B* denotes the outer grains of this region, *C* denotes the articulation between bipolar cells and cones, *D* denotes the articulation between bipolar cells and ganglion cells. In addition, *a* and *b* are the cones and rods from various regions of the retina, *e* are the bipolar cells intended for cones, *d* are the bipolar cells connected with rods, *e* are the ganglion cells, *f* are spongioblasts, *g* denotes the centrifugal fibre, *h* the optic nerve, *i* the terminate arborisations of optical fibres in geniculate bodies, *j* are the cells that receive the visual impression and *m* are the cells from which the centrifugal fibres probably originate.

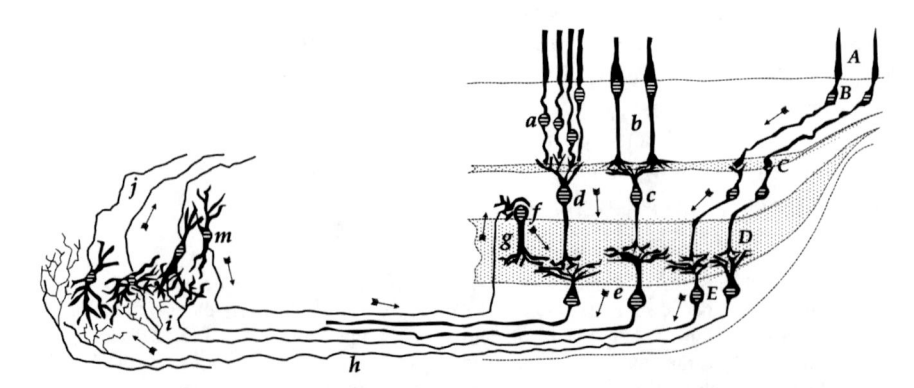

Fig. 4.48 Reproduction of Ramón y Cajal's schematic diagram for all the cells in the visual system

In the 1894 French edition *Les nouvelles idées sur la structure du système nerveux chez l'homme et chez les vertébrés* (Ramón y Cajal, 1894b), there is a whole chapter dedicate to the retina (VI.–RÉTINE). At the beginning of the chapter, Cajal makes it clear that the nervous elements of the retina are arranged in seven layers (counting the limiting membranes and the pigmentary zone),

1. Rods and cones
2. The outer grains or bodies of visual cells
3. The outer plexiform or molecular layer
4. Internal grains
5. The internal plexiform or molecular layer
6. Ganglion cells
7. Fibres of the optic nerve

All these elements are supported and isolated by large cells directed back and forth, from the outer surface of the retina to the area of cones and rods, cells that have been called Müller's fibres or retinal epithelial cells. Figure 4.49 shows a graphical representation of Cajal's figure for the retina of a mammal. These layers are apparent in this drawing which denotes *A* as the layer of cones and rods, *B* as the body of visual cells (outer grains), *C* as the outer plexiform layer, *E* as the layer of bipolar cells (internal grains), *F* as the internal plexiform layer, *G* as the layer of ganglion cells and *H* as the layer of optic nerve fibres; *a* denotes the rods, *b* the cones, *c*

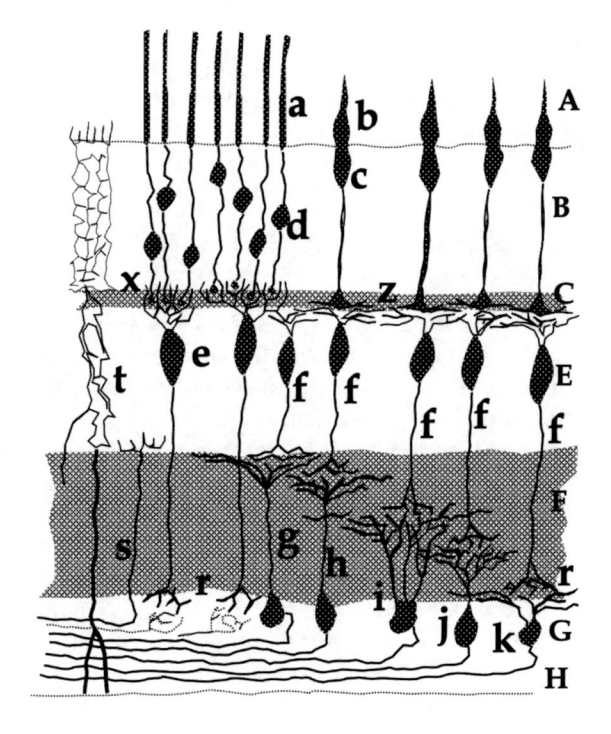

Fig. 4.49 Reproduction of Ramón y Cajal's cross section of the retina of a mammal

the body of the cone cell, d the body of the rod cell, e the bipolar cells for rods, f the bipolar cells for cones, g, h, i, j, k the branched ganglion cells in the various stages of the internal plexiform zone, r the lower arborisation of bipolar rod cells, in connection with the ganglion cells, r the lower arborisation of bipolar cells for cones, t the Müller cells or epithelial, x the contact between the rods and their bipolar cells, z the contact between the cones and their bipolar cells and s the centrifugal nerve fibre.

After presenting in detail the fine structure of the retina, Cajal proceeded with a description of its function. He was convinced that rods and cones collect light and two distinct pathways carry the information of those receptors, one for the colourless luminous intensity by the rods and one of the colour information by the cones. The visual impressions begin with the cones and rods and the signals that arise are transmitted by neural axis cylinders and are distributed by arborisations of nerve fibres. There are multiple connections among cones, bipolar cells and ganglion cells, with only an exception in the central fovea, where each cone is in contact with a single bipolar, which in turn is in contact with a limited protoplasmic arborisation of ganglionic corpuscles. Cajal connects this anatomical characteristic with the increased visual acuity in central vision. On the other hand, horizontal cells appear to bridge distinct regions of the retina including cones and rods.

Further details in the histological findings of the retina and the visual system can also be found in the 1909 and 1911 French editions *Histologie du système nerveux de l'homme & des vertébrés*, Tome I & II (*Histology of the nervous system of man & vertebrates*, Volumes I & II) (Ramón y Cajal, 1909, 1911) and particularly in Volume II.

Ramón y Cajal and Camillo Golgi (Italian biologist, 1843–1926, the inventor of the homonymous staining method) received the Nobel Prize in Physiology or Medicine in 1906.

4.22 Stephen Polyak

Stephen Lucian Polyak (1889–1955) was born Stjepan Lucian Poljak in Đurđevac (or Gjurgjevac), Austria-Hungarian Empire, later Croatia. After an adventurous early life, in 1928 he permanently moved to the United States and became a Professor of Neurology and Neuroanatomy in California and Chicago. Polyak is considered to be one of the most prominent neuroanatomists of the 20th century. Among his numerous contributions in the field, he gave a new interpretation of the basic visual processes by intensively studying the retina and revealing the role of the various cells in the retina organisation and their connectivity (Polyak, 1941, 1949, 1970). It should be noted that till that time, the prevailing theory was that rods and cones alone are responsible for colour vision (Triarhou, 2007).

Polyak was critical of the fact that theories of vision at that period focused on cones and rods alone as the building blocks of vision. He showed the retina as a complete complex instrument in which every type of cell plays its particular role in

Fig. 4.50 Reproduction of Polyak's schematic presentation of a vertical section of a human retina near the fovea

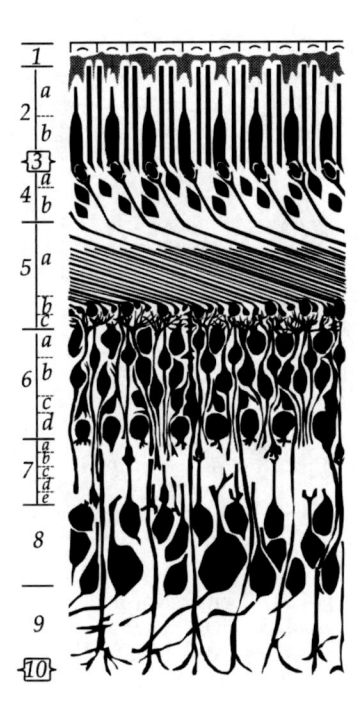

the composition of the signal that is transmitted to the brain. A brief and compact account of his theory can be found in Polyak (1949). He was particularly fond of using the Golgi staining method to attain and study the neuronal morphology of the retina. A visualisation of a part of the human retina near the fovea is shown in the Fig. 4.50, a graphical reconstruction of Polyak's original figure. The drawing reveals the basic structural elements of the human retina, which consists of layers (in numbers) and sub-layers (in letters), numbered as follows: (1) pigment layer, (2) rods and cones, (3) outer limiting membrane (4) outer nuclear layer, (5) outer plexiform layer, (6) inner nuclear layer, (7) inner plexiform layer, (8) ganglion cell layer, (9) optic nerve fibres and (10) inner limiting membrane. The top of the diagram represents the layers in the outer region of the eye. Polyak reminded that all available evidence showed that only layer (2), the rods and cones, seem to be responsible for sensing light and creating a response that propagates to the brain as nerve impulses.

Polyak further examined the central region of the retina, the central fovea, which is depicted in a graphical representation of his original figure in Fig. 4.51; in the central fovea, the cones are practically directly exposed to lights, whereas the other layers are displaced. He hypothesised that the specific structure along with the density and size of the cones in this region (and also the absence of rods) results in acute central colour vision.

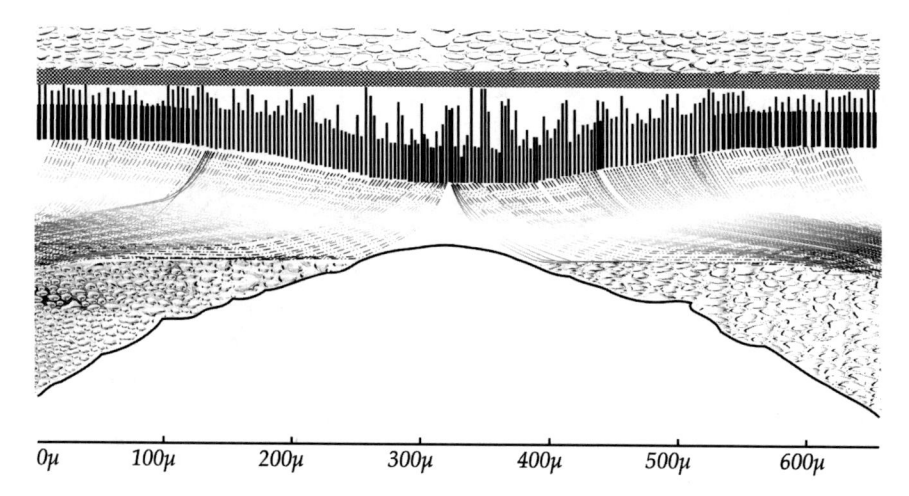

Fig. 4.51 Reproduction of Polyak's structure of the human central fovea

Fig. 4.52 Reproduction of Polyak's design of rods and cones from the central are of the Simian retina

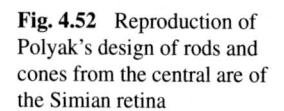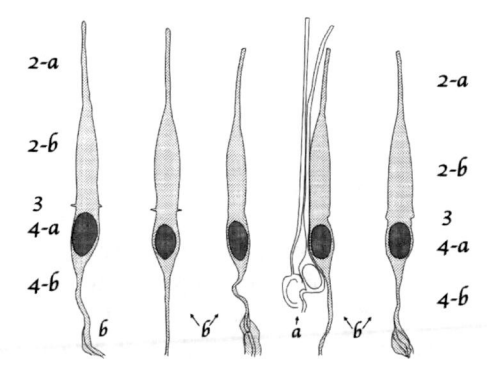

> All these various considerations force us to recognise that the only functional role of the central fovea, at any rate in Man and other Primates, is the elimination of the dioptrical impediments to the passage of the central pencil of incident light rays.

In his study, Polyak also included a figure of cones and rods alone, as graphically represented in Fig. 4.52, based on his original drawing. This is a figure of cones and rods from the central area of the Simian retina, just outside of the central (rod-less) fovea (the layer numbering is the same as in previous figures). Light enters through the bottom of this drawing and the response travels from the top. In his analysis of cones and rods, he supported, with anatomical evidence, the theory of these two totally distinct receptors.

Polyak insisted though that the other layers in the retina also play an important role in transforming the signals that travel from the receptors to the brain. He studied the bipolar cells, which he categorised into five types, and also the ganglion cells, which he also categorised into five types. He was particularly interested in understanding the interrelations of all the retinal neurons in modulating the receptors' response and understood that most theories of his time were "largely invented" (his own words). He focused his anatomical work on revealing the way the various cells connect with each other and concluded that it was time

> ...to begin to analyse and to interpret functionally the anatomical structures of the retina in terms other than rods and cones.

He affirmed the theory that cones are responsible for the perception of colour, although he was not certain of the origin of the differentiation in the spectral response that makes colour perception possible. He wondered if there was a structural or purely chemical differentiation but was not able to provide evidence from one or the other. Nevertheless, he was convinced that bipolar and ganglion cells must play an important role in colour vision.

> ...the cone – precisely because of its all-embracing universality and structural-chemical homogeneity – is not capable of performing the role of a chromatic analyser. The cone merely furnishes a dynamical 'material' for other structures of the visual system to work with.
>
> ...
>
> Plainly, the bipolar and ganglion cells must in some way be the carriers of the process by which the global cone excitation is transformed and directed into one or the other channel, according to the spectral position of the stimulus, its intensity, and other qualities.
>
> ...
>
> What arrives in the centre are the impulses originating in the cones but in many ways modified by the intervening neurons.

Polyak provided excellent illustrations of the human retina, drawings of highly educative value and built a more stable basis for further research on this rather challenging topic. In the 1950s Polyak published a highly detailed analysis of the vertebrate visual system, including facts about its origin, structure, and function, in a large volume of around 1500 pages (Polyak, 1957).

4.23   Erwin Schrödinger

Erwin Rudolf Josef Alexander Schrödinger (1887–1961) was born in Vienna, Austria-Hungary. He was a physicist whose pioneering work in physics led to the development of what is known as the *Schrödinger equation* that describes how a wave function of a system changes dynamically over time. Although he is mostly considered a significant contributor to the development of modern physics, he has done considerable work on light and colour perception. He followed the steps of the great innovators in the field of the psychology of colour perception, Newton, Helmholtz and Maxwell. Some of his published works have been translated into English and were made accessible to a wider audience, like MacAdam (1970); Niall (2017). Most of his work on light and colour vision are included in the voluptuous 1920 publications, the *Grundlinien einer Theorie der Farbenmetrik im Tagessehen* (or an *Outline of a Theory of Colour Measurement for Daylight Vision*), published in three parts (Schrödinger, 1920a, b, c). In 1925 he also published the work *Über das Verhältnis der Vierfarben zür Dreifarbentheorie*) (*About the relationship between the four-colour theory and the three-colour theory*) (Schrödinger, 1925, 1994), in which he treated the relations between the most prominent colour theories.

Schrödinger in Part I of the *Outline* (Schrödinger, 1920a) analysed topics like the concepts of light and colour, light and colour addition and the dimensionality of colour perception. Regarding the concepts of light and colour, he made clear that the normal and most precisely quantifiable way to create colours is to have light strike upon an eye. He defined colour as a group of identical-looking lights, departing from the most common designation, by which lights of the same colour can produce very different impressions under different circumstances.

> Wir entfernen uns damit ein wenig von dem gewöhnlichen Sprachgebrauch und zwar insofern, als Lichter gleicher Farbe (nach unserer Terminologie) unter verschiedenen Umständen sehr verschiedene Eindrücke auf das Auge hervorbringen konnen, so daß sie zuweilen sogar mit verschiedenen Farbnamen belegt werden.

He tried in this way to provide an objective definition of *colour*, based on its physical composition, departing from the subjective nature that may originate from the function of adaptation. In addition, he provided a physical definition for a spectral light, as a light whose function of the wavelength $f(\lambda)$ differs from zero only in a very small range of wavelengths and called the corresponding colour a *spectral colour*. Schrödinger provided a range for spectral colours, spanning roughly between 475 nm, and 630 nm in the scale of the wavelength. What is unique for these colours is that they obey his definition for $f(\lambda)$ and a group for a spectral colour includes a single member. Any mixture of lights of the same hue is expected to be less saturated (more whitish, pale) in comparison to a spectral light.

Als Spekfrallicht bezeichnen wir ein Licht, dessen $f(\lambda)$ nur in einem sehr kleinen λ-Bereich von Null verschieden ist, und die betreffende Farbe als Spekfralfarbe. Die meisten Spektralfarben (von etwa $\lambda = 475$ bis $\lambda = 630$) sind nun dadurch ausgezeichnet, daß sie sich überhaupt nur auf diese eine Art herstellen lassen —die Lichtergruppe umfaßt nur das eine Licht— wenn man davon absieht, daß bei genügend klein gewähltem λ-Bereich die Verteilung der Energie innerhalb dises Bereiches willkürlich ist, weil hinreichend benachbarte Wellenlängen sich in ihrer Wirkung auf das Auge nicht unterscheiden. Es gibt wohl Mischlichter vom gleichen Farbton, sie, erscheiden aber gegen das Spektrallicht immer etwas weißlich (weniger gesättigt).

Regarding the *dimensionality* of colour vision, Schrödinger made three basic statements that clearly support the trichromacy theory, but also emphasise the subjective nature of colour perception, as the dimensionality varies for three distinct cases.

- For normal colour vision (trichromats): there are linearly independent colour triples; four colours are always linearly dependent.
- For partially colour-blind people (dichromates): there are linearly independent pairs of colours; three colours are always linearly dependent.
- For totally colour-blind people (monochromats): every two colours are linearly dependent.

A. Für Farbentuchtige (Trichromaten): Es gibt linear unabhängige Farbentripel. Vier Farben sind stets linear abhängig.
B. Für partiell Farbenblinde (Dichromaten): Es gibt linear unabhängige Farbenpaar. Drei Farben sind stets linear abhängig.
C. Für total Farbenblinde (Monochromaten): Je zwei Farben sind linear abhängig. Diese Aussagen bedeuten nichts anderes, als daß die Farbenmannigfaltigkeit für diese Personen drei bzw. zwei.

He supported the tridimensionality of the colour space manifold by pointing out that (a) any two spectral colours differ in more than one quantity, their intensity, in a way that any attempt to match them only by changing the intensity is insufficient, thus pairs of spectral colours are linearly independent; (b) any binary mixtures of spectral colours cannot be matched by any other single spectral colour with changes in intensity only, thus triplets are linearly independent.

In Part II of the *Outline* (Schrödinger, 1920b), Schrödinger focused on the representation of the colour space manifold, which he called the *envelope of the spectral cone*. A reproduction of his 1920 graph is shown in Fig. 4.53. The graph shows the spectral cone as a curve (*Spektralkurve*) and its intersection with an inclined plane (denoted as *Schnitt mit einer Ebene*). In this representation, he commented on the

Fig. 4.53 Reproduction of
Schrödinger's envelope of
the spectral cone

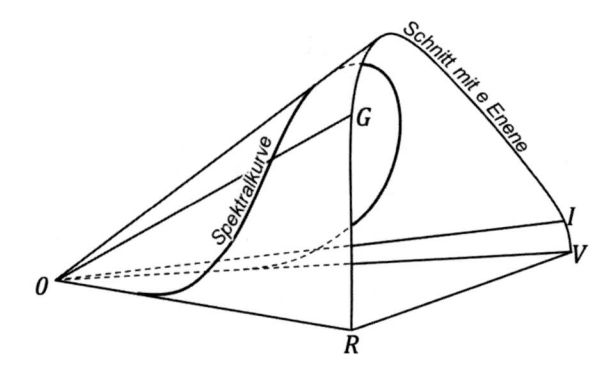

prominent position of white, which is the colour of sunlight. On one hand, he diminished its significance by stating that it is purely the radiation of a black body of about 7000° C, and on the other, he supported its significance, as the evolutionary reason for colour vision. It is no surprise that the solar radiation peaks in the range of light that is visible. This is a very important statement that shown how the understanding of the development and the function of vision started to get a more clear 'shape' at the beginning of the 20th century.

Denn fast ausschließlich unter der Einwirkung dieses Lichtgemisches ist unser Auge entstanden, hat es sich entwickelt und seine gegenwärtige Funktionsform angenommen. Daß dieses Lichtgemisch für die so entstandene Farbwahrnehmung eine ausgezeichnete, zwischen den möglichen Extremen vermittelnde Rolle spielt, ist nicht verwunderlich; ist doch, nebenbei bemerkt, auch für die merkwürdige Koinzidenz des Energiemaximums der Sonne mit dem Helligkeitsmaximum in einem Spektrum von konstanter Energie die einzige ungezwungene Erklärung die phylogenetische; an der Stelle des Energiemaximums und zu beiden Seiten desselben war die Entwicklung hoher Lichtempfindlichkeit sozusagen am rentabelsten, wenn es auf möglichst deutliche Wahrnehmung der Gegenstände auch bei schwacher Beleuchtung ankam.

Furthermore, Schrödinger in an attempt to achieve a theoretical determination of the coordinate system for the colour space, provided another important graph, that of colour-mixture curves of the interference (uniformly dispersed) spectrum of sunlight, in relative coordinates across the visible spectrum. This corresponds to the case in which light is determined by only its numerical function of wavelength $(f(\lambda))$. By using trichromacy and by trying to calculate the colour vector to be assigned to sunlight white purely mathematically, he produced these curves, a reproduction of which is provided in Fig. 4.54; the graph shows the normalised colour-mixture curves of the spectrum of sunlight with respect to three primary

Fig. 4.54 Reproduction of Schrödinger's colour-mixture curves of the interference spectrum of sunlight, in relative coordinates across the visible spectrum

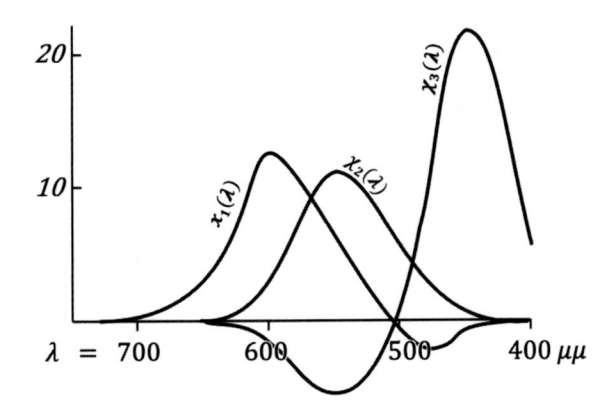

colours (or calibration lights), which he arbitrarily chose to be the red at the far end of the spectrum, the green at $\lambda = 505$ nm, and violet at the other far end of the visible spectrum. Since the idea was to produce the sunlight white, the proportion of these three colour-mixture curves is meant to produce white (the colour qualitatively resembling sunlight).[32] The curves are normalised in such a way that they have the same area between them and the abscissa.

In Schrödinger's analysis, supposing light as a function of wavelength $f(\lambda)$ then the three colour coordinates of that light (given a reference light $\phi(\lambda)$) can be expressed as

$$\int \frac{f(\lambda)x_i(\lambda)}{\phi(\lambda)}\,d\lambda, \quad i = 1, 2, 3 \tag{4.15}$$

where $i = 1, 2, 3$ are the three dimensions of the colour space, x_i is the ith colour mixture function. Schrödinger remarked that only the relative illumination is necessary (f/ϕ) for the computations. For purely spectral colour light sources this ratio is 1, thus, the coordinates that result are just the averages of each of the three integrals of the colour mixture functions.

$$\frac{1}{C} \int x_i(\lambda)\,d\lambda, \quad i = 1, 2, 3 \tag{4.16}$$

where C is the width of the wavelength interval.

In Part III of the *Outline* (Schrödinger, 1920c), Schrödinger provided advanced colour measurement concepts using strict mathematical notation, including measures of perceptual dissimilarity, an analysis of the concept of brightness and the sensitivity to colour dissimilarity, the introduction of the concept of geodesics in the estimation of colour dissimilarities on his colour space. In his understanding, *colour theories emerge as expressions of concepts that attempt to agree with the sensation.*

---

[32] This graph was derived from the *elementary sensation curves* published by A. König and C. Dieterici, Zeitschrift für Psychologie und Physiologie der Sinnesorgane 4., P. 241ff., 1892.

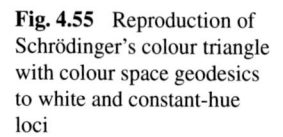

Fig. 4.55 Reproduction of
Schrödinger's colour triangle
with colour space geodesics
to white and constant-hue
loci

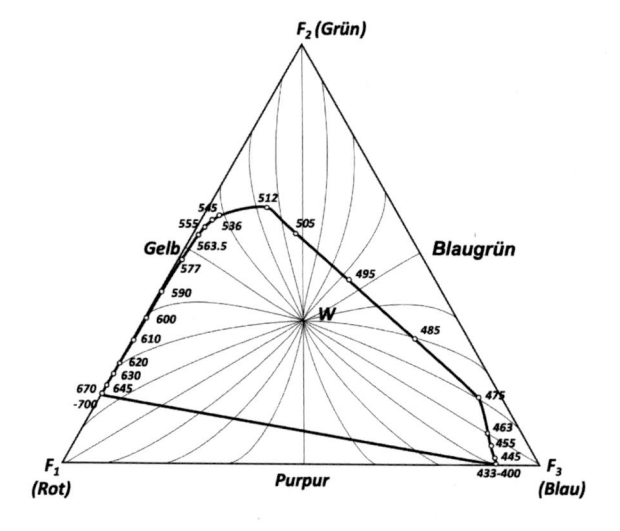

...ich betone ausdrücklich, daß es sich dabei nicht um eine "Fol-
gerung aus der Theorie" handelt, die über Empfindungen natürlich
überhaupt keine Aussage macht, sondern nur um den Versuch,
einen Begriff, dessen bisherige quantitative Fassung zu äußerlich
ist und sich mit der Empfindung nicht deckt, anders und tiefer zu
fassen und so, daß er sich womöglich besser mit der Empfindung
decke.

He stated that the colours that lie on the shortest path to white are equal in hue.

Als farbtongleich mit einer gegebener Farbe hätte man dann also
jene Farben zu definieren, die auf der kürzesten Linie zum Weiß
liegen.

Figure 4.55 is a reproduction of the 1920 graph (Fig. 6) that Schrödinger provided
to depict a colour triangle and the pencil of geodetic lines through the white point, in
which colours of the same brightness are considered. In this representation, only the
connections to the primary colours and their complementaries are rectilinear, and
only in these cases would the mixture with white be constant in hue. For all intermediate colours, the hues of the mixtures with white would appear shifted towards
the predominant primary colour. A colour near the white point has its equivalent
in hue along the edge of the triangle, which can be located by progressing along a
curve of the pencil.

Figure 6 zeigt das Büschel geodätischer Linien durch den
Weißpunkt in einem Farbendreieck, in welchem wir uns gleich-

helle Farben lokalisiert denken wollen. Geradlinig sind nur die Verbindungen zu den Grundfarben und ihren Gegenfarben, nur in diesen Fällen wären die Weißmischungen konstant im Farbton. Von allen Zwischenfarben erscheinen die Weißmischungen im Farbton verschoben gegen die überwiegende Grundfarbe, d.h. gegen diejenige Grundfarbe, welche darin —verglichen mit dem Mischungsverhältnis der drei Grundfarben, das farblos erscheint— am stärksten vertreten ist. Denn man findet ja nach Annahme zu einer Farbe in der Nähe des Weißpunkts die farbtongleichen am Rande des Dreiecks, indem man längs einer Kurve des Büschels fortschreitet, und das Büschel zieht sich in einem dreistrahligen Stern gegen die Grundfarbenpunkte zusammen.

4.24 Ludwig Wittgenstein

Ludwig Wittgenstein (1889-1951) was born in Vienna, Austria-Hungary who worked in logic, the philosophy of mathematics, language and the mind. His most important publication was the *Tractatus Logico-Philosophicus* (*Treatise on Logic and Philosophy*), published in 1921 in a form of a collection of quotes (Wittgenstein, 1922). Wittgenstein was concerned with accurate symbolism like it is expressed in the relation of one fact (such as a sentence) to another so that it can be a symbol for that other. By analysing language and logic and by exploring the limitations of human thinking, *Wittgenstein defined language to be the limit to the expression of any thought*. His teacher, the world-known mathematician and philosopher Bertrand Russell (1872–1970) described Wittgenstein *as perhaps the most perfect example of genius*, thus it is more than important to know of Wittgenstein's research in his own words.

Wittgenstein used the term 'picture' (or image) in a broader sense for mappings in general, although the consequences in human perception insofar this treatise is concerned are important to understand. In paragraphs 2.1–2.1512 of the original text in German[33] he states that we make to ourselves pictures of facts. The picture presents the facts in logical space, the existence and non-existence of atomic facts. The picture is a model of reality. To the objects correspond in the picture the elements of the picture. The elements of the picture stand, in the picture, for the objects. The picture consists in the fact that its elements are combined with one another in a definite way. The picture is a fact. That the elements of the picture are combined with one another in a definite way, represents that the things are so combined with one another. This connection of the elements of the picture is called its structure, and the possibility of this structure is called the form of representation of the picture. The form of representation is the possibility that the things are combined with

---

[33] The complete text with the English translation can be found online @ https://archive.org/details/tractatuslogicop1971witt.

one another as are the elements of the picture. Thus the picture is linked with reality; it reaches up to it. It is like a scale applied to reality.

Wir machen uns Bilder der Tatsachen. Das Bildstellt die Sach-
lage im logischen Raume, das Bestehen und Nichtbestehen von
Sachverhalten vor. Das Bild ist ein Modell der Wirklichkeit. Den
Gegenstaenden entsprechen im Bilde die Elemente des Bildes.
Die Elemente des Bildes vertreten im Bild die Gegenstande. Das
Bild besteht darin, dass sich seine Elemente in bestimmter Art und
Weise zu einander verhalten. Das Bild ist eine Tatsache. Dass sich
die Elemente des Bildes in bestimmter Art und Weise zu einander
verhalten stellt vor, dass sich die Sachen so zu einander verhal-
ten. Dieser Zusammenhang der Elemente des Bildes heisse seine
Struktur und ihre Moeglichkeit seine Form der Abbildung. Die
Form der Abbildung ist die Moeglichkeit, dass sich die Dinge so
zu einander verhalten, wie die Elemente des Bildes. Das Bild ist
so mit der Wirklichkeit verknuepft; es reicht bis zu ihr. Es its wie
ein Masstab an die Wirklichkeit angelegt.

In paragraphs 2.161–2.171 of the original text Wittgenstein states in the picture and the pictured there must be something identical so that the one can be a picture of the other at all. What the picture must have in common with reality in order to be able to represent it after its manner-rightly or falsely-is its form of representation. The picture can represent every reality whose form it has. The spatial picture, everything spatial, the coloured, everything coloured, etc.

In Bild und Abgebildetem muss etwas identisch sein, damit das
eine ueberhaupt ein Bild des anderen sein kann. Was das Bild
mit der Wirklichkeit gemein haben muss, um sie auf seine Art
und Weise richtig Oder falschabbilden zu koennen, ist seine Form
der Abbildung. Das Bild kann jede Wirklichkeit abbilden, deren
Form es hat. Das raeumliche Bild alles Raeumliche, das farbige
alles Farbige, etc.

Among the publications of Wittgenstein that were edited and published posthumously was a collection of *Bemerkungen über die Farben* (*Remarks on Colour*) (Wittgenstein, 1977), which is of particular interest for this treatise. In paragraph I.22, he states that we do not want to establish a theory of colour (neither a physiological one nor a psychological one), but rather the logic of colour concepts. And this accomplishes what people have often unjustly expected of a theory.

Wir wollen keine Theorie der Farben finder (weder eine physiolo-
gische, noch sine psychologische), sondern die Logik der Farbbe-
griffe. Und diese leistet, was man sick oft mit Unrecht von einer
Theorie erwartet hat.

It is clear in this collection of remarks that Wittgenstein was familiar with other
colour theories, like the one Goethe derived, and also knew Runge's ideas about
colour. The text, nevertheless, largely reflects his own way of philosophical thinking. Wittgenstein seems to have been intrigued by 'transparent' colours, the colour
of transparent media like coloured glasses. He was positive that coloured glasses
work like filters, taking away the colours of any objects behind them and changing their appearance according to their own colour. He was particularly concerned
about white and argued that *there could not be a transparent white*. He asserted
that *opaqueness is not a property of the white colour*. He seems to have accepted a
four-colour model based on blue, yellow, red and green. In Part III of the Remarks,
Wittgenstein poses some interesting questions and answers regarding the knowledge and perception of colour. For example, in III.4 he wonders whether a red is
lighter than a blue and if this is a matter of experience. He shows a deep understanding of concepts like saturation in saying (in III.14) that a saturated X colour
is an impression of colour in a particular surrounding and comparable to its 'transparent' version. In paragraph III.52 he states that we only need six colour words
to communicate about colours (which supports his adoption of the four+two colour
model). In the same paragraph, he adopts Runge's view on not being possible to
communicate using self-opposing colour phrases, like reddish-green or yellowish-
blue. At the same time, in III.61, he marks a core question to always have in mind,
which asks "how do people learn the meaning of colour names". Later in the text,
in III.106 he reminds that "the logic of the concept of colour is just much more
complicated than it might seem". His philosophical inclination to the study of this
subject is emphatically illustrated in paragraph III.120, where he remarks on the
case of colour blindness,

Do normally sighted people and colour-blind people have the
same concept of colour-blindness? And yet the colour-blind person understands the statement "I am colour-blind", and its negation as well. A colour-blind person not merely can't learn to use
our colour words, he can't learn to use the word colour-blind
exactly as a normal person does. He cannot for example always
determine colour-blindness in cases where the normal-sighted
can.

4.25 Gestalt Psychology

In any historical account of colour theory, a note should also be made about a particular movement that came from a mixture of psychological and physiological studies,

the movement of *Gestalt* theory.[34] Gestalt is a significant paradigm shift in psychology, born in Germany at the beginning of the 20th century. It was *Christian von Ehrenfels* (1859–1932), an Austrian philosopher, who gave this movement its name in *Über Gestaltqualitäten* (*The Attributes of Form*), his most important work, published in 1890 (Ehrenfels, 1890). The term Gestalt cannot be literally translated into English and is usually interpreted as 'structure', or 'totality', or 'configuration', or even 'organised unity'. Gestalt proposed a paradigm shift to the dominant psychology of that period and made significant contributions to cognitive psychology, which was built around the idea that *the whole is more than the sum of its parts*. This phrase seems to originate in Aristotle's Μετά τα φυσικά (*Metaphysics*), in which[35] is stated that

> Περὶ δὲ τῆς ἀπορίας τῆς εἰρημένης περί τε τοὺς ὁρισμοὺς καὶ περὶ τοὺς ἀριθμούς, τί αἴτιον τοῦ ἓν εἶναι; πάντων γὰρ ὅσα πλείω μέρη ἔχει καὶ μὴ ἔστιν οἷον σωρὸς τὸ πᾶν ἀλλ' ἔστι τι τὸ ὅλον παρὰ τὰ μόρια, ἔστι τι αἴτιον, ἐπεὶ καὶ ἐν τοῖς σώμασι τοῖς μὲν ἁφὴ αἰτία τοῦ ἓν εἶναι τοῖς δὲ γλισχρότης ἤ τι πάθος ἕτερον τοιοῦτον.

This translates to

> With regard to the difficulty which we have described in connexion with definitions and numbers, what is the cause of the unification? *In all things which have a plurality of parts, and which are not a total aggregate but a whole of some sort distinct from the parts, there is some cause*; inasmuch as even in bodies sometimes contact is the cause of their unity, and sometimes viscosity or some other such quality.

Among the main characteristics in the Gestalt theory, one may distinguish

- The complexity of the human mind cannot be reduced.
- Mental representations do not correspond completely with those that exist in reality; people construct them by themselves.
- Through perception, people are able to acquire knowledge of the world, interact with it and connect with others.
- Gestalt theory focuses on visual perception.

---

[34] Among numerous sources about Gestalt theory there is a specific website dedicated to it @ http://www.gestalttheory.net/ and belongs to the *International Society for Gestalt Theory and its applications*.

[35] The ancient Greek text and the English translation of 1933 by Hugh Tredennick was used, found online @ https://archive.org/details/in.ernet.dli.2015.185284/mode/2up, Book 8, (par. 1045a.1), par. VI, pp. 420–423.

According to Gestalt, people tend to mentally connect and integrate contiguous objects, connect elements to a single object or group, fill the gaps in shapes, perceive similar objects as having the same form by abstraction, and simplify the representations.

Historically, it is generally accepted that there are three founders of Gestalt, *Wolfgang Köhler, Max Wertheimer* and *Kurt Koffka. Wolfgang Köhler's* (1887–1967) main contribution was the formal introduction of *learning by discovery* and the basic notion that this process is active and dynamic (Köhler, 1925, 1847).

The key phrase of Gestalt (the whole is more than the sum of its parts) is attributed to him. The *phi-phenomenon* or apparent movement is *Max Wertheimer's* (1860–1943) most revolutionary discovery (Wertheimer, 1912). Simply described, according to this phenomenon, movement is perceived when a succession of images is presented, the basic concept for cinema. His most important works in Gestalt are included in the two volumes of the *Untersuchungen zur Lehre von der Gestalt* (*Investigations on Gestalt Theory*) (Wertheimer, 1923, 1922), in which he outlines the general theoretical context and the laws of organisation in perceptual forms. *Kurt Koffka* (1886–1941) contributed in a variety of fields, including memory, learning and perception. He also applied Gestalt to fields such as child psychology (Koffka, 1924). He emphasised the need to consider mental processes from a holistic point of view. He also helped Wertheimer in his research on the apparent movement by becoming involved as a subject. His main contribution to the understanding of Gestalt is included in his 1935 *Principles of Gestalt Psychology* Koffka (1935).

Gestalt soon became a new *school of thought*, which viewed human behaviour and perception as a complete whole. It led to major contributions in explaining some complex processes of sensation and, particularly, perception, focusing on the notion that humans perceive the world by viewing things in totality or from a holistic perspective.

4.26 John Guild

John Guild was a British scientist who worked at the National Physical Laboratory (NPL) in Teddington, England. His main work was in the development of a wide variety of optical instruments and techniques (Carter, 1991). Guild's contribution to the 20th century (and onward) colour technology is significant, and his work has been discussed in various publications and presentations by other researchers (Carter, 1991; Fairman et al., 1997). Two of his papers can be found in MacAdam's collection of colour science sources (MacAdam, 1970), namely *Some problems of visual perception* Guild (1970b) and *Quantitative data in visual problems* (Guild, 1970a).

Guild's contribution was mostly marked by his work towards the formulation of a common standard for colour space representation, the *CIE 1931* standard.[36] The relevant work along with tables with his measurements can be found in his 1931 publication on *The colorimetric property of the spectrum* (Guild, 1931), which he presented as work he had done two-three years before publication, and in which he included and integrated more experiments conducted by William David Wright (1929). A detailed description of the CIE colourimetric standard was published by him and Thomas Smith (1931), including all the functions, data and resolutions adopted. Guild was the NPL Representative in the CIE discussions towards the standardisation and served at the CIE Secretariat Committee on Colourimetry for 1928–1931.

Motivation for Guild's work was the need for standardisation in colour science at the beginning of the 20th century; this included the definition of the *standard observer*, the *standard human vision fundamentals* and *colour matching functions*, along with the *standard colour space* for a variety of applications. He was seriously involved in what was known as the colour matching experiments, in order to derive a concrete and mathematically expressed colour space for an objective description of colours. In these experiments, he used the standard white light defined by the National Physical Laboratory (the NPL Standard White Light). The spectral distribution of this light is shown in Fig. 4.56, reproduced from Guild's Table I in

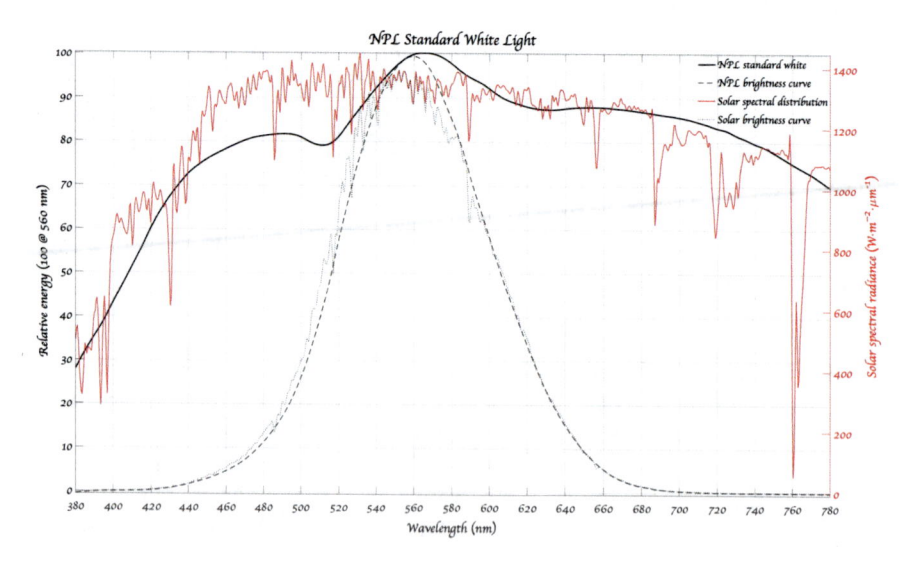

Fig. 4.56 Energy distribution of the NPL Standard White Light

---

[36] CIE stands for Commission Internationale de l'Echlairage (International Commission on Illumination). As stated on CIE's website, the commission "is devoted to worldwide cooperation and the exchange of information on all matters relating to the science and art of light and lighting, colour and vision, photobiology and image technology.".

Guild (1931)[37]; the luminous efficacy data (the normalised response of the eye to luminance) and the perceived brightness of the white light are also shown from the original data Guild provided, along with the solar radiation energy distribution on the Earth's surface and the perceived brightness of the Sun, from modern high-resolution data. As it was already known, one simply needs to multiply the spectral distribution of a light source by the luminous efficacy of the eye to get the perceived brightness.

The goal of the original experiments was to create mixtures of three primary spectral (monochromatic) colours in order to match a given colour and derive the human vision colour matching functions. Supposing a linear colour system (adopting Grassmann's laws) in which any spectral colour Q_λ (or, as Guild named it, *colourimetric quality of a monochromatic stimulus*) is a linear combination of primaries R, G, B,[38]

$$Q_\lambda = aR + bG + cB \qquad (4.17)$$

where R, G, B are the primaries, which Guild set to 700, 546.1 and 435.8 nm, the NPL standard primaries,[39] perceived as red (at the limit of vision), green and violet-blue respectively, as shown in an approximate representation in Fig. 4.57[40]; a, b, c are the trichromatic coefficients, which may be estimated by getting the contribution of each primary guided by instrument adjustment factors and the convention that $a + b + c = 1$.

Overall, the setup of the experiment and the outcomes is graphically shown in Fig. 4.58 and the details will be laid out in the following paragraphs.

First, the experiment takes into account each observer's sensation of white, which becomes the reference for the adjustment of the experimentation instrument (this is some equivalent to white matching or white balancing). For example, if for an observer the white is attained by 30 units of R, 60 units of G and 20 units of B, then, since only equal amounts of primaries should be mixed to attain white, R should be scaled by $a_w = 2$, G by $b_w = 1$ and B by $c_w = 3$. These factors characterise the colour matching of the specific observer, thus they are applied to any colour matching experiment that the observer participates in. If, for example, the

---

[37] Solar radiation data were adopted from The National Renewable Energy Laboratory of the U.S. Department of Energy. The '2000 ASTM Standard Extraterrestrial Spectrum Reference E-490-00' where found at https://www.nrel.gov/grid/solar-resource/spectra-astm-e490.html and directly downloaded through https://www.nrel.gov/grid/solar-resource/assets/data/e490_00a_amo.xls. The 'Reference Air Mass 1.5 Spectra' data were found at https://www.nrel.gov/grid/solar-resource/spectra-am1.5.html and directly downloaded through https://www.nrel.gov/grid/solar-resource/assets/data/astmg173.xls.

[38] By that time, R, G, B was the standard representation of the set of the three primary colours, even though it was known that any other independent colour triplets could be used interchangeably.

[39] In subsequent experiments, Guild transformed the standard primaries into *working primaries* by using (4.17) and the results of the initial experiments, and by considering R, G, B to be the working primaries and Q_λ the standard primaries. Inverting this system of linear equations he attained the working primaries for the subsequent experiments.

[40] Any triangle that connects primaries encloses the complete gamut of colours that can be produced by those primaries assuming the additive system. The white triangle is showing the locations of the NPL primaries. For comparison, the much later standard showing the sRGB triangle is also drawn in the diagram.

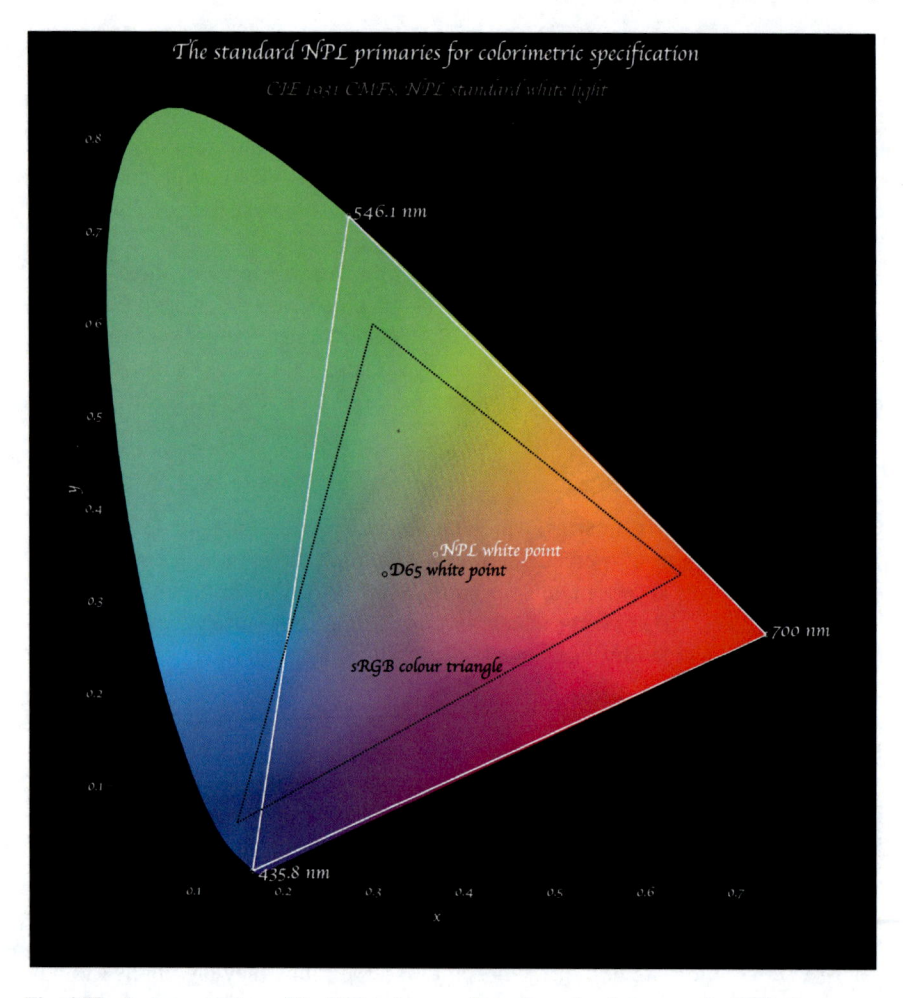

Fig. 4.57 A representation of the NPL primary colours for colourimetry on the CIE colour space that Guild helped develop

observer matches a colour with {40, 20, 20} units for the primaries, then these values are scaled with the factors {2, 1, 3} and become {80, 20, 60} which add to 160; taking the appropriate ratios, the coefficients are estimated as $a = 80/160 = 0.5$, $b = 20/160 = 0.125$ and $c = 60/160 = 0.375$. It is easy to understand that for any triplet that corresponds to a different colour Q_λ, different coefficients are estimated. These measurements describe the *chrominance* of a coloured light but say nothing about its *luminance*. This is why he also included *luminosity matching* in the experiments, which completed the picture about the information extracted by the visual system for a coloured light, by providing three more parameters, the *luminosity factors* L_R, L_G and L_B. By white balancing the chromaticities and using the three luminosity factors he was able to create a single luminosity factor L_λ. To arrive at a

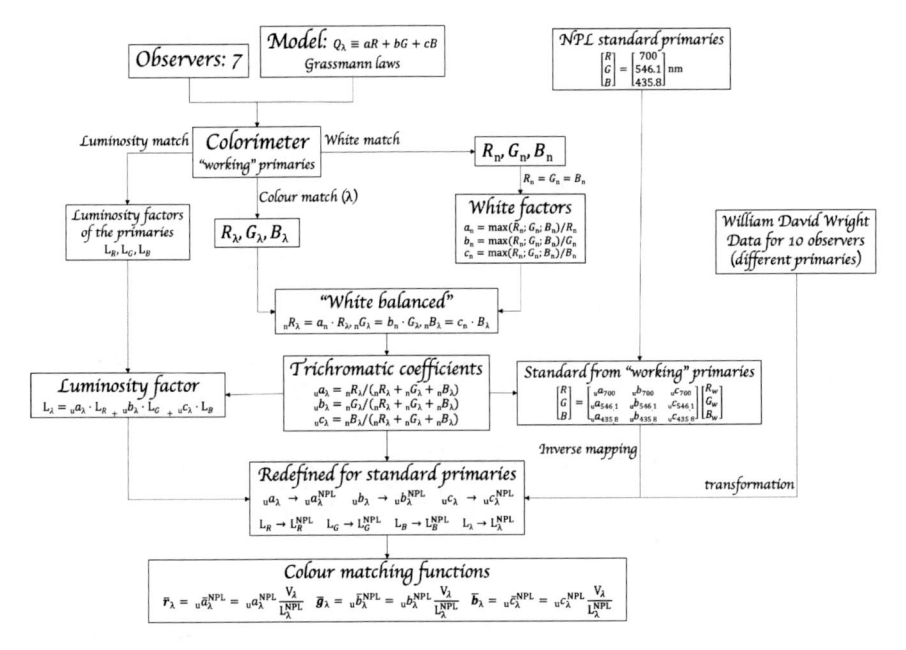

Fig. 4.58   Graphical representation of Guild's experiment

result that would be universal and independent of the three primary colours, he compensated his results for the NPL standard primaries, since he knew the primaries in his colourimeter were different. Thus he arrived at the unit functions of wavelength for the chrominance and luminosity factors $_u a_\lambda^{NPL}$, $_u b_\lambda^{NPL}$, $_u c_\lambda^{NPL}$ and L_λ^{NPL}, such that

$$Q_\lambda = {}_u a_\lambda^{NPL} R + {}_u b_\lambda^{NPL} G + {}_u c_\lambda^{NPL} B$$
$$_u L_\lambda = {}_u a_\lambda^{NPL} L_R + {}_u b_\lambda^{NPL} L_G + {}_u c_\lambda^{NPL} L_B \qquad (4.18)$$
$$_u a_\lambda^{NPL} + {}_u b_\lambda^{NPL} + {}_u c_\lambda^{NPL} = 1$$

After experimenting with filters, he concluded to the luminosity factors $L_R = 1.0$, $L_G = 2.858$, $L_B = 0.169$, to be used as relative factors. Then he recalculated the values to compensate for inconsistencies in the results and concluded to $L_R = 1.0$, $L_G = 4.39$, $L_B = 0.048$,[41] for which he included the results for the standard observer in Table IV.

By repeating the experiment for the whole range of the visible spectrum, the whole variety of the coefficients was exposed, giving rise to specific response functions. Guild did that experiment with seven observers and incorporated the experimental results by Wright (1929) for another ten observers and represented the average results in tables and graphs. The average results give specific values for the *normal eye*, the standard observer and how the colour specification should be carried out. In Fig. 4.59 these results are shown in terms of three graphs, which were created

---

[41] These luminosity factors of the primaries are expressed as a ratio and not as absolute values; they should be considered as $L_R : L_G : L_B = 1 : 4.39 : 0.048$.

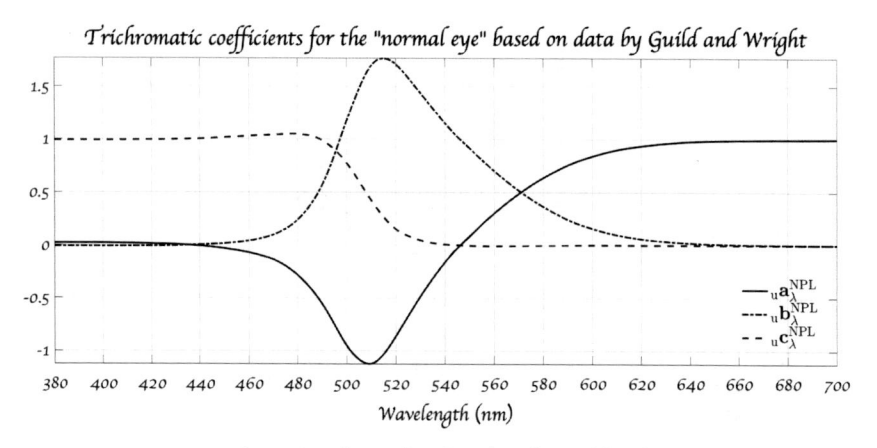

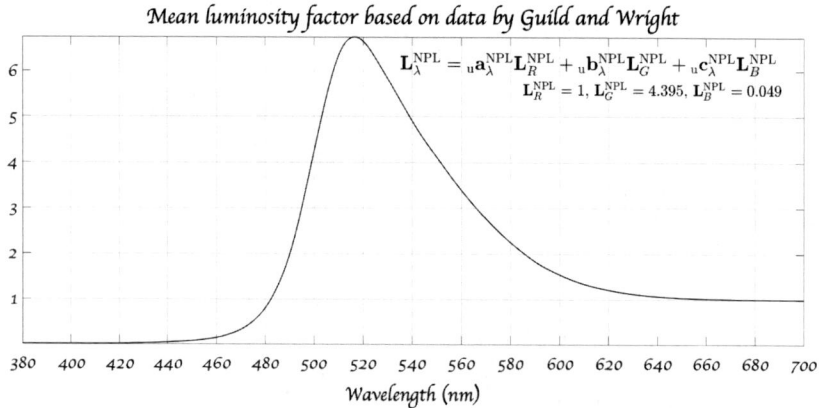

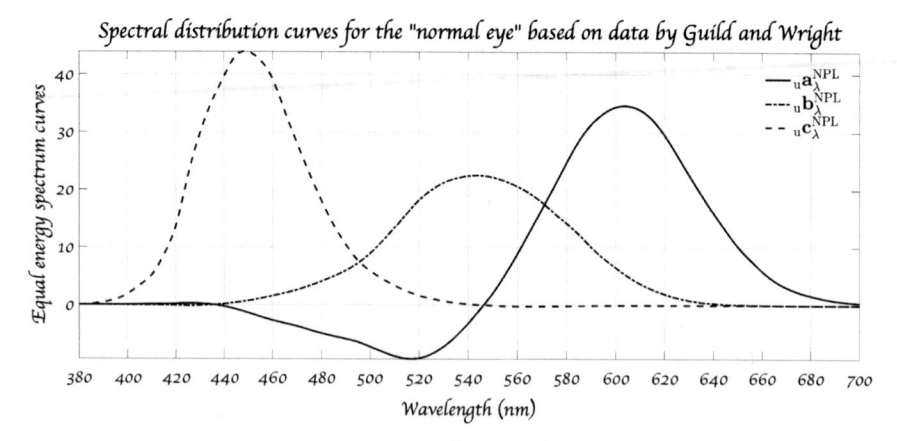

Fig. 4.59   Specification of the standard observer by the Guild-Wright experiments

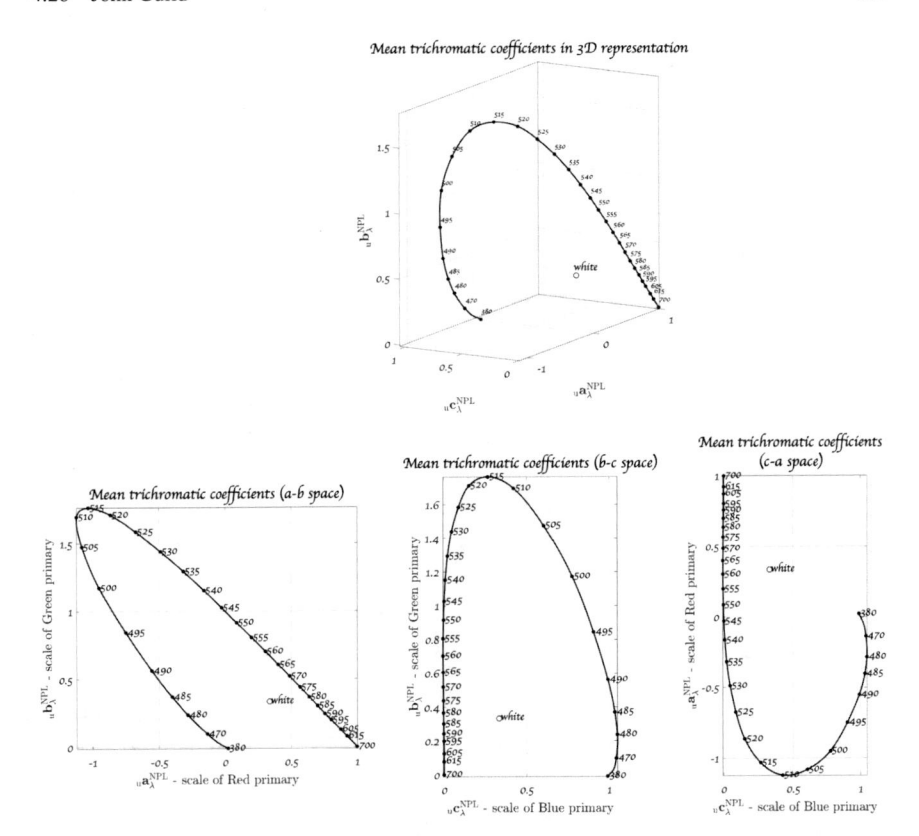

Fig. 4.60 Chromaticity diagrams by the Guild-Wright data

by interpolating the values from the original data. The first graph is, for the first time, the picture of the standard observer, which shows the standard trichromatic coefficients of the *normal eye* ($_u a_\lambda^{NPL}$, $_u b_\lambda^{NPL}$, $_u c_\lambda^{NPL}$), whereas the second graph shows the mean luminosity factor (L_λ^{NPL}) for the complete visible spectrum. The third graph shows the spectral distribution curves of the primaries for the standard observer that are produced by assuming an equal energy spectrum of an arbitrary value of 100 at all wavelengths.

Figure 4.60 show the resulting chromaticity graphs from the Guild-Wright experiments, regarding the standard observer and for the primaries at 700, 546.1 and 435.8 nm (NPL white light). The three-dimensional graph on the top shows the actual shape of the chromaticity space in three dimensions. The three two-dimensional graphs show the same space in the possible two-coefficient combinations. The graphs only show the spectral colours, marked by their corresponding wavelength and the location of the white. All colours are within the concave space that the spectral curves denote and by imagining a line connecting the 380 nm end with the 700 nm end (the so-called line of purples).

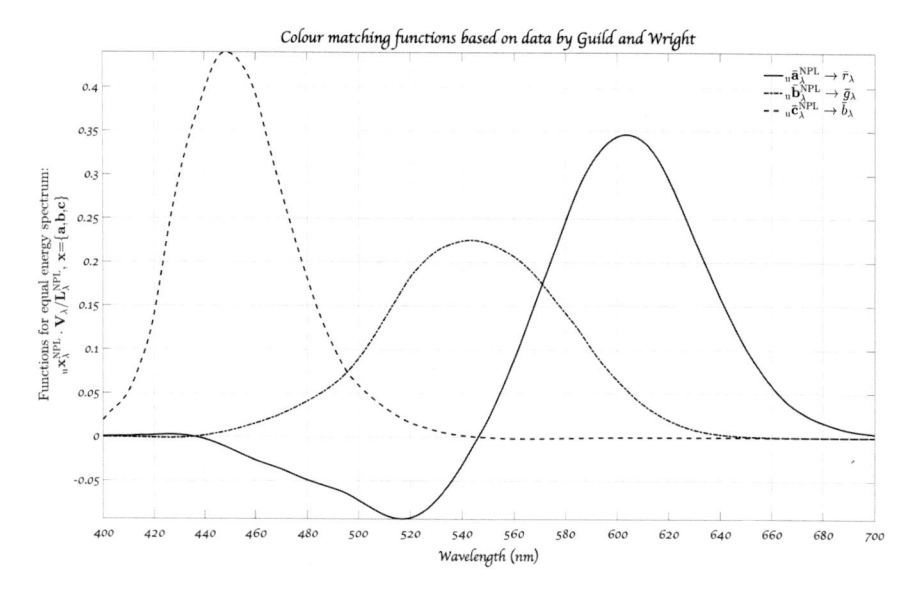

Fig. 4.61   Colour matching functions for the NPL primaries by the Guild-Wright data

The most useful colour matching functions can be produced by multiplying the trichromatic coefficients by the ratio of the luminous efficacy (shown in Fig. 4.56) to the luminosity factor,

$$\bar{r}_\lambda = {}_u\bar{a}_\lambda^{\text{NPL}} = {}_ua_\lambda^{\text{NPL}}\frac{V_\lambda}{{}_uL_\lambda^{NPL}}$$

$$\bar{g}_\lambda = {}_u\bar{b}_\lambda^{\text{NPL}} = {}_ub_\lambda^{\text{NPL}}\frac{V_\lambda}{{}_uL_\lambda^{NPL}} \hspace{2cm} (4.19)$$

$$\bar{b}_\lambda = {}_u\bar{c}_\lambda^{\text{NPL}} = {}_uc_\lambda^{\text{NPL}}\frac{V_\lambda}{{}_uL_\lambda^{NPL}}$$

The colour matching functions that result are shown in Fig. 4.61.

In a paper most probably published some time after the establishment of the CIE 1931 standard, Guild with T. Smith (both members of the CIE Committee during that period) laid out a more concrete and corrected version of his original data and publication (Smith & Guild, 1931). Smith and Guild presented the general context, the corrected data and the resolutions included in the standard. Initially, they distinguished between the notion of colour as twofold; on one side it is to be considered *a subjective visual sensation* and on the other *an objective attribute of a physical stimulus* and its colour-matching relation to other stimuli. In addition, they presented a definitive view of the basic colour theory at that time, which stated that *colours are expressed in a linear trichromatic system using three primaries that correspond to*

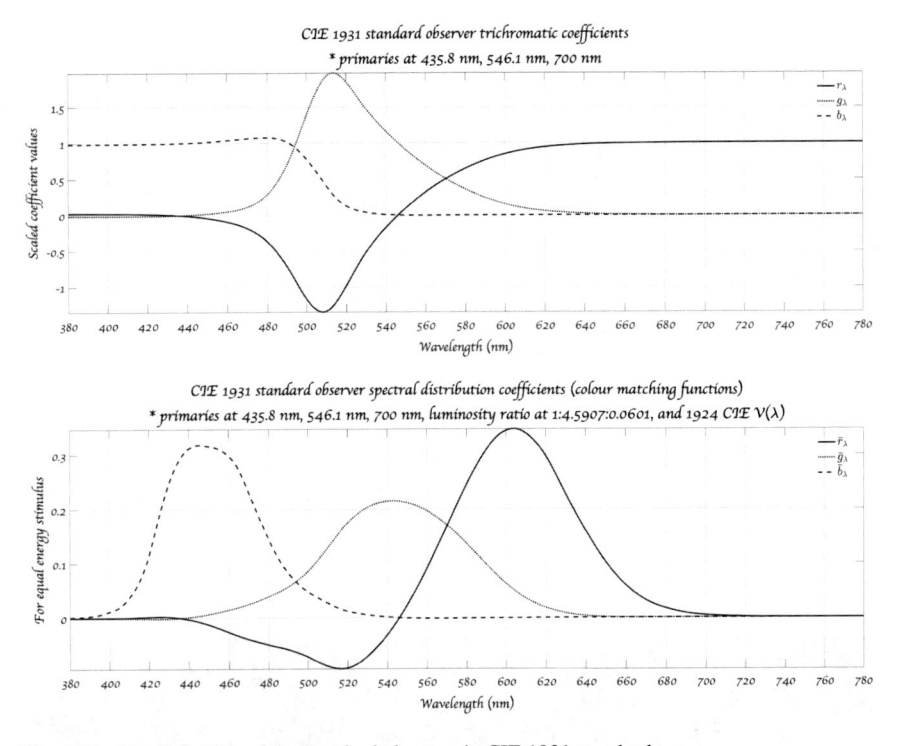

Fig. 4.62   The definition of the standard observer in CIE 1931 standard

linearly independent colour sensations. The presentation of the CIE 1931 standard was based on the description of the *five resolutions* that constituted it.

Resolution 1 defined the notion of the standard observer as defined by Guild's previously presented data and slightly corrected for better accuracy. Those data defined the CIE rgb colour space[42] (as trichromatic coefficients and colour matching functions) and the corresponding data were included in Table I of Smith & Guild (1931). The NPL standard white light was used as a reference with primaries at 435.8, 546.1 and 700 nm and the relative luminosities were fixed at the ratio of 1:4.5907:0.0601. Figure 4.62 shows a graphical representation of the data by Smith and Guild regarding the CIE 1931 standard observer. The first graph shows the r_λ, g_λ, b_λ trichromatic coefficients, whereas the second graphs shows the colour matching functions $\bar{r}_\lambda$, $\bar{g}_\lambda$, $\bar{b}_\lambda$.

Resolution 2 defined the standard illuminants to be used for colour applications. These illuminants were simply referenced as *A*, *B* and *C* and corresponded to a gas lamp at 2848 K for reference *A*, the same lamp with liquid filters composed of particular copper and cobalt sulphate solutions for reference *B*, and the same lamp with different liquid filters again based on copper and cobalt sulphate solu-

---

[42] The red-green-blue (rgb) are used by convention to correspond to the perceived colours of the used primaries. Other primaries could easily result in another colour space.

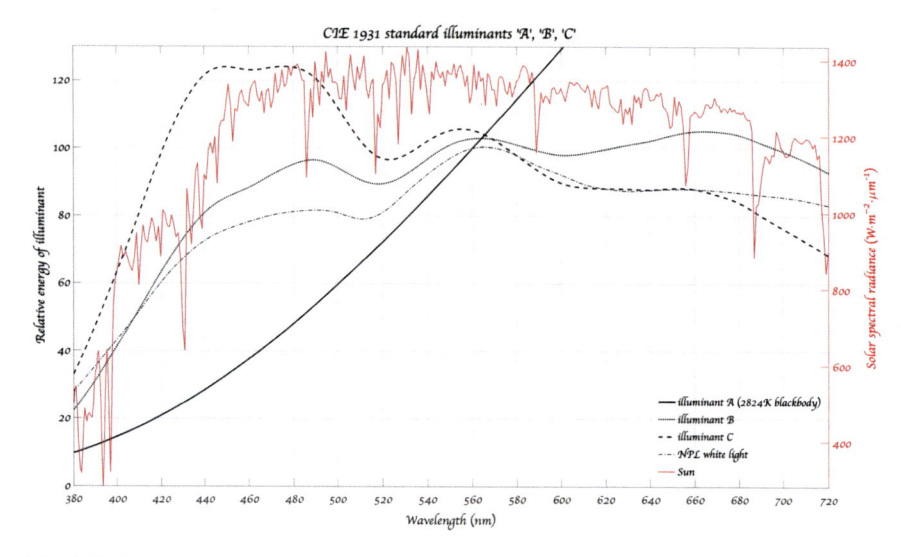

Fig. 4.63   The spectral power distribution of the CIE 1931 standard illuminants

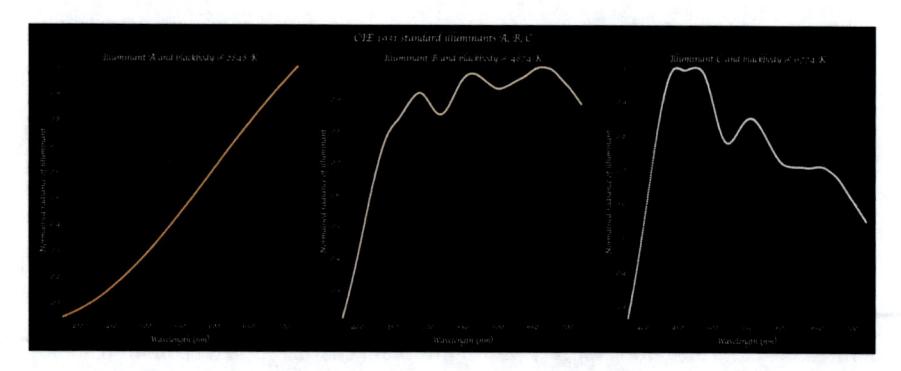

Fig. 4.64   Approximation of the radiation of the standard illuminants by black body radiation

tions for reference *C*. Tables II and V of Smith & Guild (1931) included the relative energy spectral distribution of the illuminants for practical applications. In the analysis presented in the paper, it was specified that the illuminant *B* corresponds to a blackbody at a temperature of 4800 K (currently estimated at 4874 K) and that the illuminant *C* corresponds to a blackbody at 6500 K (currently estimated at 6774 K). In addition, the previously used NPL standard white light was mostly similar to illuminant *C*. Figure 4.63 depicts the relative spectral power distribution of the three standard illuminants in the visible spectrum. For comparison, the graph includes the distribution of the NPL standard white light and the solar radiation at sea level (right-side y-axis). In addition, Fig. 4.64 shows the spectral power distribution of the standard illuminants in comparison to approximating blackbody radiation, depicted in a colour corresponding to the matching colour temperature.

Resolution 3 defined the conditions and setup for the colourimetric measurements of reflective and opaque media.

Resolution 4 defined that colourimetry is based on trichromacy and any colour should be expressed as a linear combination of three independent scales.

Resolution 5 defined the CIE 1931 XYZ colour space, as a projection of the original RGB space to a space where (a) no negative values are allowed and (b) one of the three variables corresponds to the perceived luminance. To fulfil these conditions, the X, Y and Z variables correspond to *fictional primaries*, totally outside the perceivable colour gamut. This process is presented both in terms of an analysis of the rationale and the method and is supported by detailed data in Tables III and IV of Smith & Guild (1931). This is one of the most analysed parts of the work done for the CIE 1931 standard. Originally, Smith & Guild (1931) presented this projection using a two-dimensional representation of the colour space, and particularly the blue-green representation. Smith and Guild emphasised that the estimates for the definition of the new x, y, z axes were not strict, as only the *alychne* was a mathematically and conceptually established notion. As defined by Schrödinger (1994), the alychne is the line of zero perceived luminance. Using the *luminosity factor* defined in (4.18), this translates to having

$$r + g\mathrm{L}_G + b\mathrm{L}_B = 0 \qquad (4.20)$$

in which the luminosity factors of the primaries are normalised by the factor of the first primary ('R' by convention), corresponding to 1:4.5907:0.0601 (as stated in Resolution 1). As the colour matching function for the second primary ('G' by convention) is extremely close to the general luminosity sensation (luminous efficacy), the variable to be used for the side closer to that primary was Y coordinates. Thus the alychne should include the X and Z coordinates. Figure 4.65 graphically depicts the $b - g$ colour space, as Smith and Guild also did, and the transformation to the $z - y$ colour space, whereas Fig. 4.66 presents the resulting $\bar{x}, \bar{y}, \bar{z}$ colour matching functions based on the estimate of that xyz space, as described in the following paragraphs.

To solve this for the case of the $b - g$ space that Smith and Guild used, the equation of the XZ line is estimated by

$$s_{XZ} = \frac{\mathrm{L}_R - \mathrm{L}_B}{\mathrm{L}_G - \mathrm{L}_R}$$
$$k = \frac{(r_0 - 1)\mathrm{L}_R + g_0\mathrm{L}_G + b_0\mathrm{L}_B}{\mathrm{L}_G - \mathrm{L}_R} \qquad (4.21)$$
$$g = s \cdot b + k$$

where s_{XZ} is the slope of the line, k is the displacement factor and (r_0, g_0, b_0) are the chromaticity coordinates of the B primary, which corresponds, according to the Guild data, to $(0.0272, -0.0115, 0.9843)$. To determine the XY and ZY line equations there is no strict method according to Smith and Guild. They just reference

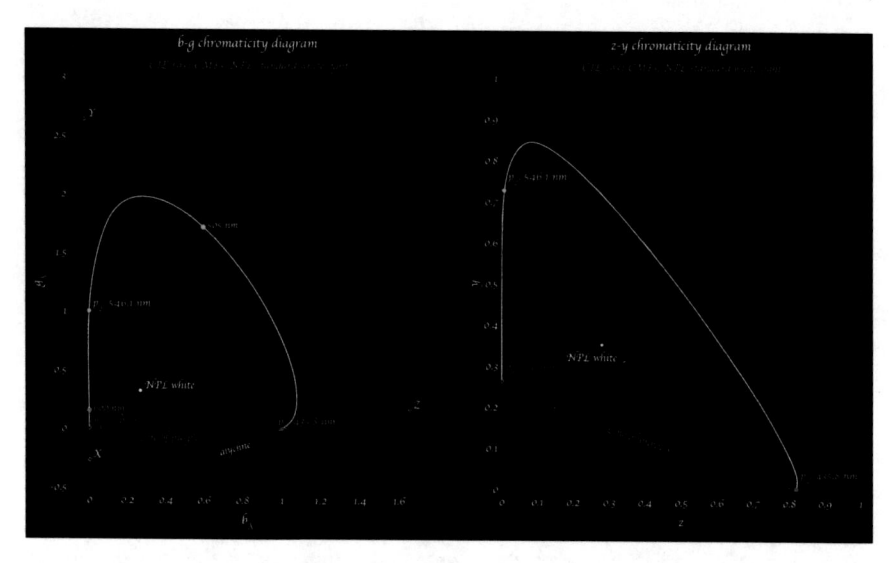

Fig. 4.65   Derivation of the $z - y$ colour space from the $b - g$ colour space

CIE 1931 standard observer spectral distribution coefficients XYZ (colour matching functions)

*primaries at 700 nm, 546.1 nm, 435.8 nm, luminosity ratio at 1:4.5907:0.0601, and 1924 CIE V(λ)

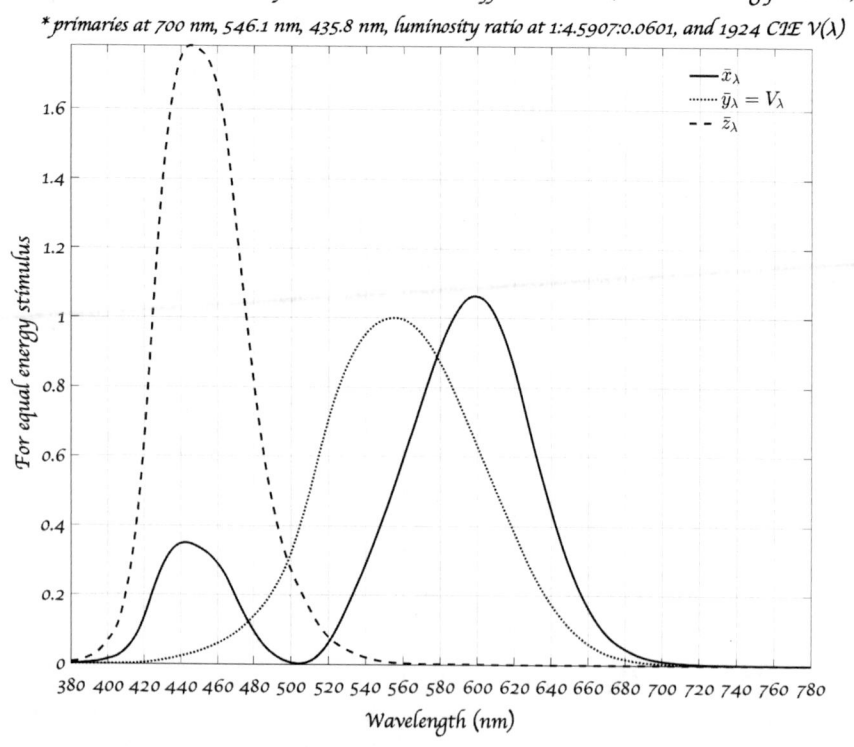

Fig. 4.66   Estimated colour matching functions from Smith and Guild data

suitable selections that fit the needs of applications and still fulfil the two basic prerequisites stated above. They state that the XY line should be tangent to the spectral locus at the location of the R primary. Thus it is difficult to proceed and replicate their results. Having also in mind the fact that the Tables they presented are relatively sparse and there is noise in the measurements, one may resort to interpolation techniques and suggestions by other researchers (Fairman et al., 1997; Wold & Valberg, 1999; Service, 2016). For example, Fairman et al. (1997) suggest to find the tangent to the spectral locus at the position of 670 nm. In the case of the experiments of the author, the interpolated original data where still noisy at that location and thus the 600 nm location was selected. The corresponding XY line equation can be expressed using the derivative of the locus at the selected location,

$$s_{XY} = \frac{dg_{[600nm]}}{db_{[600nm]}}$$
$$g = s_{XY} \cdot b \tag{4.22}$$

s_{XY} is the slope of the XY line. The same strategy is used to find the ZY line, which, according to Smith and Guild *should pass near the spectral locus in a direction that secures satisfactory metrical properties.* Again following Fairman et al. (1997), the selected locus point is at 505 nm and the corresponding ZY line equation can be expressed using the derivative of the locus at the selected location,

$$s_{ZY} = \frac{dg_{[505nm]}}{db_{[505nm]}}$$
$$g = s_{ZY} \cdot (b - b_{[505nm]}) + g_{[505nm]} \tag{4.23}$$

where s_{ZY} is the slope of the ZY line. The intersection of the three lines result the locations of the X, Y and Z primaries in the RGB space as

$$XYZ = \begin{bmatrix} 1.2661 & -1.6116 & -0.81979 \\ -0.26852 & 2.6351 & 0.16421 \\ 0.0023934 & -0.023487 & 1.6556 \end{bmatrix} \tag{4.24}$$

where the columns of the XYZ matrix are the coordinates X, Y and Z primaries in the RGB space. To transform to the fictional XYZ space assuming unitary vectors XY, XZ, ZY one should simply find the inverse of XYZ matrix and normalise it. This results a transformation matrix T which can be used to project the RGB colour space to the new XYZ space.

$$T = \begin{bmatrix} 0.4879 & 0.30029 & 0.21181 \\ 0.17413 & 0.82109 & 0.0047813 \\ 0.0000 & 0.0088345 & 0.99117 \end{bmatrix} \tag{4.25}$$

The transformation matrix that Smith and Guild presented is slightly different due to their different approximation.

$$T = \begin{bmatrix} 0.49000 & 0.31000 & 0.20000 \\ 0.17697 & 0.81240 & 0.01063 \\ 0.00000 & 0.01000 & 0.99000 \end{bmatrix} \tag{4.26}$$

The transformation from RGB to XYZ space can be based on the formula suggested by Fairman et al. (1997),

$$
rgb = \begin{bmatrix} r_\lambda \\ g_\lambda \\ b_\lambda \end{bmatrix}
$$

$$
s = \sum_{i=1,2,3} T_{i,1} \cdot r_\lambda + \sum_{i=1,2,3} T_{i,2} \cdot g_\lambda + \sum_{i=1,2,3} T_{i,3} \cdot b_\lambda
$$

$$
x = T_1 \cdot rgb \oslash s
$$
$$
y = T_2 \cdot rgb \oslash s
$$
$$
z = T_3 \cdot rgb \oslash s
$$
$$\tag{4.27}$$

where $T_{i,j}$ is any i, j element of the transformation matrix and T_i is any i row vector of the matrix. The $\oslash$ operator denotes the element-wise division. Applying this transformation to the RGB data Smith and Guild provided the XYZ data obtained are shown in the $x - y$ diagram in Fig. 4.65. A more familiar and common representation is that of the $r - g$ space transformation to $x - y$ space, as shown in Fig. 4.67, following the same principle as in the case of the $b - g$ space. Smith and Guild provided extensive data in Tables III and IV in 1 nm intervals.

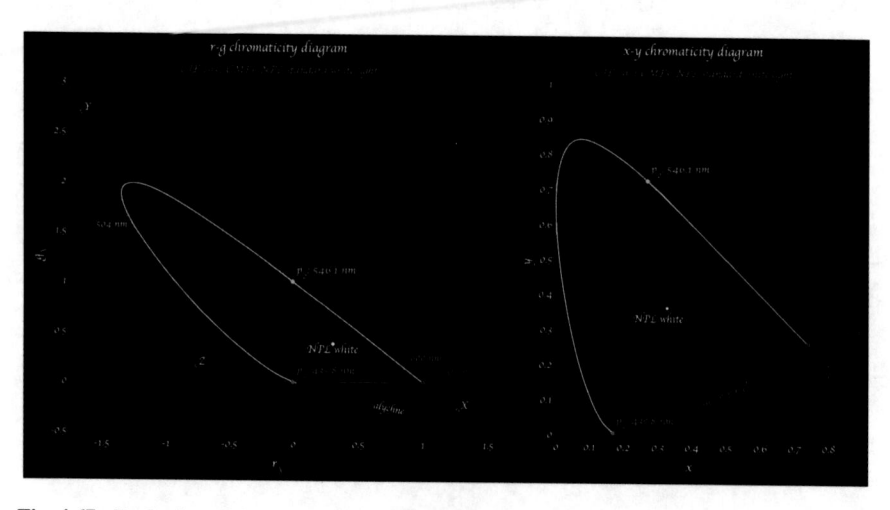

Fig. 4.67   Derivation of the $x - y$ colour space from the $r - g$ colour space

Guild in his 1970 work *Some problems of visual perception* found in MacAdam's collection (Guild, 1970b), presented a critical view of the current approaches and even the semantics of colourimetry and visual perception theories. He emphatically clarified that *in any colour matching experiment, the observer need never operate more than three independent controls, as, in general, three are necessary, and in no circumstances are more than three required.* Then he tried to also clarify the stages involved in a process of radiation reception, in general, which also applies to human vision. Thus he recognised four stages, including three objectives (1–3) and one subjective (4).

- Stage 1: the receptor, which interacts directly with radiation and determines the spectral sensitivity of the overall sensing system
- Stage 2: the coupler, which is the interface that transmits intensity-based messages to the next stages
- Stage 3: the indicator, which transforms the initial detection event message to a representation adequate for a higher level recognition
- Stage 4: the cognizer, which is the part of the system that *becomes aware* of the state of stimulation of the receptor.

By applying this general concept to vision, Guild concluded that *there are three, and only three, independent reception systems in simultaneous operation in human vision.* Consequently, he formulated a general hypothesis that *there could be any number, not less than three, of different types of receptor, and there could be central connections of any number of different modalities, not less than three.* He also proposed three other alternative theories, by which either the number of modalities or the number of the receptors may vary in any combination of those. Furthermore, he analysed those hypotheses in terms of logic based on evolutionary biology to consider some of them more improbable than others, but due to lack of knowledge in the mechanics of the human vision at the receptor level, he was unable to support one or another.

Chapter 5
Epilogue

—Die Welt ist meine Vorstellung: dies ist die Wahrheit, welche in Beziehung auf jedes lebende und erkennende Wesen gilt...Die Welt ist Vorstellung.

Arthur Schopenhauer, Die Welt als Wille und Vorstellung

Following Thomas Young's chronology of optical authors (Young, 1807c), the timeline of the history presented in this treatise is shown in Fig. 5.1. It is a story of almost two and a half millennia, written by protagonists of a variety of characters, full of love and passion, success and failure, joy and sorrow, peace and fight. In the era of philosophy, colour science was founded upon logic, experience and philosophical worldviews. The middle ages marked a shift towards theorisation with experimental verification, still influenced by personal and societal worldviews. A significant paradigm change in the relevant research occurred only after the 16th century when thinkers adopted an investigation approach towards objective truth, supported by new theories, advanced technology and ingenious experimentation.

In a journey through time from the 6th century BCE to the mid 20th century, this treatise aimed at revisiting the ideas of the pioneers in visual perception in an attempt to identify the foundations of colour science. The text depends heavily on in-depth research for the original texts in various resources so that the newer generations be in contact with the old writings and the ideas be brought into the spotlight using the original words and sketches of the pioneers. Original texts have been presented accompanied with translations to contemporary language, as this was imperative due to changes in the usage and meaning of many words and expressions of notions and ideas in the past, since the context changed, in some cases, significantly.

From Alcmaeon's vision through the water and fire of the eyes to Guild's colour science standardisation, the human genius needed twenty-five centuries to reach a usable and verifiable theory for colour science. The foundations of this science were

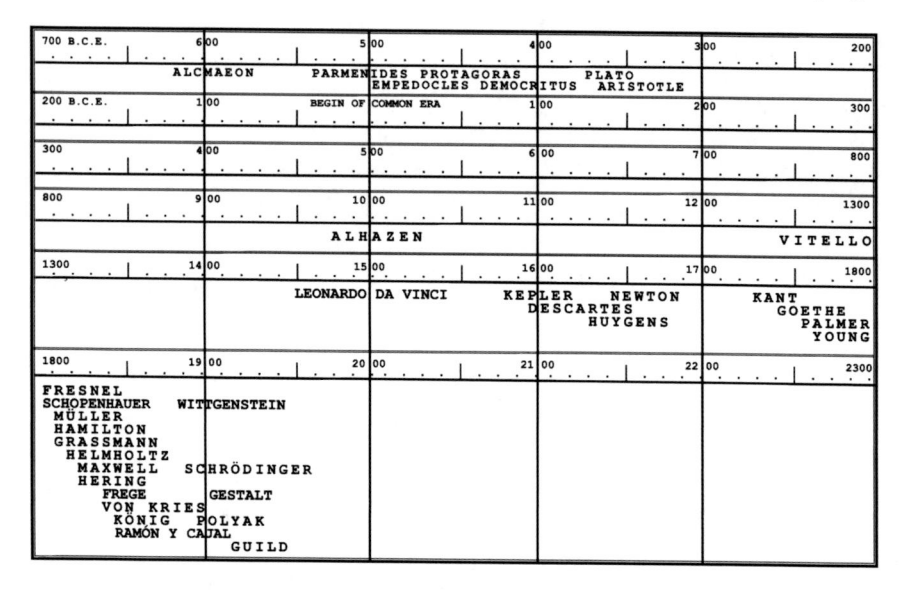

Fig. 5.1 A timeline for the foundations of colour science

built by cross-disciplinary research efforts in physiology, neurology and anatomy, physics, mathematics and signal processing, linguistics and philosophy. Still, those foundations, continue to be under investigation and revisiting, as no definitive and conclusive theories provide a unified theory for colour science. That the nature of light is currently dual, might be an indication of an incomplete theory for its description. That the mechanism of colour sense, sensation and perception is currently suggested to be based on complex electrochemical interactions in one of the most energy-consuming parts of the brain, still under intense investigation, is another indication of the breakthroughs to expect in the future. Even the very basic photo sensing cells of the eyes, the cones and rods, which seem to act as the low-level sensors of light, have not been conclusively modelled as regards the capacity to sense and differentiate light wavelengths (there is also a waveguide theory that departs from the most accepted three-cone-type theory). That various technologies that apply the 'standardised' colour science still exhibit serious limitations, is yet another indication of the need for technological improvements. That technologies to provide artificial colour vision still evolve to increase the human capacity in terms of resolution and spectral bandwidth, sketches another aspect of advancements.

What is currently known is that the perception of colour is based on a complex light-sensing process in the eyes and a further process in the brain. The two types of photoreceptors in the eyes, the cones and rods, host special types of opsins that are light-sensitive proteins due to the chromophore retinal, which is a form of Vitamin A, and is the chemical basis of animal vision. When photons enter the eyes and hit the cones, they trigger a transformation of their retinal that begins the *chemical signalling cascade* to result in colour perception. The type of opsin bound to the retinal defines the range of wavelengths that are absorbed, thus different opsin-retinal combinations result in different absorbance spectra. This difference in the absorbed spectra is labelled by the brain using specific types of labels, the colours.

It was the year 1967 when the Nobel Prize in Physiology or Medicine was awarded jointly to Ragnar Granit, Haldan Keffer Hartline and George Wald "for their discoveries concerning the primary physiological and chemical visual processes in the eye".[1] George Wald was the first to discover that Vitamin A was essential in retinal function. During the 1950s, Wald and his colleagues were able to extract pigments from the retina and conduct spectrophotometer measurements of light absorbance on the pigments. George Wald was the key person to formulate the modern theory of *visual phototransduction*, by which light is transformed into an electrical signal in the eyes.

In human eyes, there are four types of photoreceptor proteins (opsins) that are responsible for vision

- *Rhodopsin*, which is found in rod cells and enables night colourless vision, with a maximum response at a wavelength of around 500 nm (in the region of what is perceived as green in daylight vision).
- *Long-wavelength sensitive opsin*, with a maximum response at a wavelength of around 560 nm, which is perceived as a yellow-green colour. Despite the yellow-green appearance at the peek response, this opsin is called the *red opsin*, or *erythrolabe*, of more generically *L opsin*, or *LWS opsin*. The naming convention by which this opsin is labelled as *red* comes from the fact that it is the opsin most sensitive to the red part of the spectrum.
- *Middle-wavelength sensitive opsin*, with a maximum response at around 530 nm, which corresponds to the perception of green colour, and thus, called the *green opsin*, or *chlorolabe*, of more generically *M opsin*, or *MWS opsin*.
- *Short-wavelength sensitive opsin*, with a maximum response at around 430 nm, which is perceived as a blue colour, and thus, called the *blue opsin*, or *cyanolabe*, or more generically *S opsin*, or *SWS opsin*.

Figure 5.2 shows a reproduction of the spectral sensitivities of the human cone pigments estimated from psychophysics based on data from (Stockman & Sharpe, 2000).[2]

Nevertheless, colour perception is more than just the chemical signalling cascade or visual phototransduction in the cones and rods, as more complex phenomena take place, like that of the adaptation in lighting conditions and scene composition, the colour opponency imposed by the ganglion cells, the compression of the neural signal and its transmission to the brain, and the role of memory, experience and possibly language in the labelling and perception of colour in an image.

---

[1] The Nobel Foundation, 1967, The Nobel Prize in Physiology or Medicine 1967, online at https://www.nobelprize.org/nobel_prizes/medicine/laureates/1967/.

[2] The data were found online at http://www.cvrl.org/pigments.htm. It should be noted that although the various studies resulted in slightly different results, the general idea holds and the regions in all studies agree.

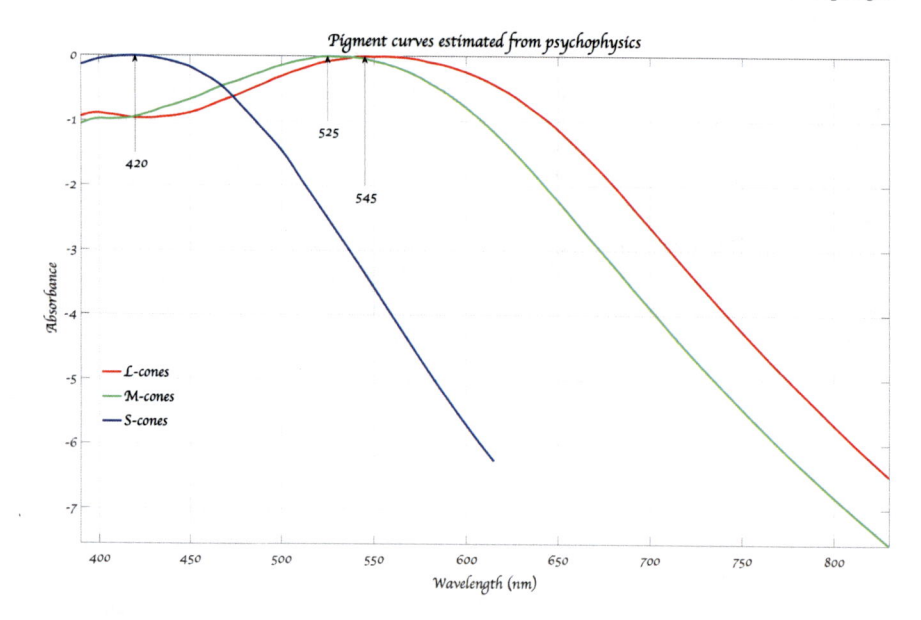

Fig. 5.2 Cones' photopigment absorption

An objective colour science, sterilised from an observer, would be a colourless radiation science about light and optics. This is clearly evident in the work of many of the great minds whose contributions have been included in this treatise, those particularly involved in physics. They preferred to focus their research on light, its propagation and properties and the devices to manipulate it, with no reference whatsoever to notions of colour. Those of them that tried to explain colour, were able to do so only in a form of average experimental results in ingeniously designed experiments. Only by the integration of the work of neuroanatomists was it possible to reach an understanding of the processes underlying the very definition of the notion of colour. The ever-growing technological development formalised colour science and it now has a mostly technical aspect, in many cases forgetting the subjective character of colour. Colour systems have been largely standardised, as has been the cross-platform colour rendering, in an attempt to create stimuli that are interpreted as colours in the minds of the technology users. Advanced methods and algorithms drew on colour science to make image exchange over distances possible. What would the Web be without JPEG and MPEG[3]? Colour science is interwoven with information science at the core of the technology to enable image and video streaming.

In any case, colour theory is a theory about colour, which is a purely subjective phenomenon that becomes 'objective' only as an average collective sensation. Colour exists only in the minds of living beings, in a process that evolved as a perception

---

[3] JPEG and MPEG are the world-known image and video compression methods that deeply exploited colour science theory to attain vast amounts of data compression and made high-quality image and video interchange possible.

mechanism, most probably to increase the survival chances. In the advanced human civilisation, colour perception became much more than a survival mechanism, as it became a means for expression and pleasure, an integral part of the arts and aesthetics. Objectively, colour is the end-product of a conscious mind that labels the various wavelengths of light to differentiate among the objects and their properties in a scene. In this view, colours can be regarded as names or as labels to different quantised 'qualities' of light or 'qualities' of the objects. Protagoras' words still sound relevant, *man is the measure of all things.*

References

Alhazen. (1021). *Book of optics* (1572 edn.). Basile Ae, Per Episcopios.

Archer-Hind, R. D. (1888). *The timaeus of plato*. London, UK: Macmillan and Co.

Aristotle. (1993). *Metaphysics*. Kaktos. ISBN: 9789603522218, (Original work published ca. 350 BCE).

Aristotle. (1994). *Meteorologica 2 (iii, iv)*. Kaktos. ISBN: 9789603522256, (Original work published ca.350 BCE).

Aristotle. (2001). *On the soul*. Georgiadis—Library of the Greeks. ISBN: 9789603161219 (Original work published ca.350 BCE).

Aristotle. (2004). *On sense and the sensible*. Zitros, 9789604631148, (Original work published ca.350 BCE).

Benson, J. (2000). *Greek colour theory and the four elements*. UMass Amherst Libraries.

Burchardt, J. (2004). The dispersion of sunrays into colours in crystal by Witelo. *Organon, 33*, 69–82.

Carter, E. C. (1991). In M. Halstead & J. Holmes (Eds.), *The advancement of colour, a history of the first fifty years of the colour group (great britain) 1941–1991*. London: Colour Group of Great Britain.

Clegg, B. (2002). *Light years and time travel: An exploration of mankind's enduring fascination with light*. Wiley.

Codellas, P. S. (1932). Alcmaeon of croton: His life, work, and fragments. *Proceedings of the Royal Society of Medicine, 25*(7), 1041–1046. https://www.ncbi.nlm.nih.gov/pmc/articles/PMC2183733/.

Davidson, T. (1869). *The fragments of parmenides* (Vol. Translated into English hexameters with Introduction and Notes). St. Louis, USA: E.P. Gray.

Descartes, R. (1637a). In J. Translation: Veitch, (Ed.), *Discourse on the method of rightly conducting the reason and seeking truth in the sciences*. Chicago, 1903: Kegan Paul, Trench, Trübner & Co., Ltd.

Descartes, R. (2001). *1637b*. Electronic Edition by Jean-Marie Tremblay: La dioptrique.

Descartes, R. (1637c). In J. Editor: Beck C.H.; Translation: Veitch, (Eds.), *Selections from the principles of philosophy*. Munich, 1901: Ezreads Publications LLC.

Descartes, R. (1644). *Principia philosophiae*. Amstelodami, apud Ludovicum Elzevirium.

Descartes, R. (1664). *De homine figuris*. Lugduni Batavorum: Ex officinâ Hackiana.

© The Editor(s) (if applicable) and The Author(s), under exclusive license to Springer Nature Singapore Pte Ltd. 2021
G. Pavlidis, *A Brief History of Colour Theory*,
https://doi.org/10.1007/978-3-030-87771-2

Descartes, R. (1667). *Discours de la methode: Pour bien conduire sa raison, & chercher la verité dans les sciences: Plus la dioptrique, et les meteores. Qui sont des essais de cette methode.* Paris: Chez Theodore Girard.

Descartes, R. (1677). *Tractatus de homine.* Apud Danielem Elsevirium.

Diels, H., & Kranz, W. (1903). *Die fragmente der vorsokratiker griechish und deutsch.* Berlin: Weidmannsche buchhandlung.

Edgar, S. (2015). The physiology of the sense organs and early neo-kantian conceptions of objectivity: Helmholtz, lange, liebmann. In F. Padovani, A. Richardson, & J. Y. Tsou (Eds.), *Objectivity in science—New perspectives from science and technology studies* (ISBN 978-3-319-14348-4 ed., pp. 101–122). Cham: Springer. https://doi.org/10.1007/978-3-319-14349-1_6.

Ehrenfels, C. V. (1890). über gestaltqualitäten. *Vierteljahrsschrift für wissenschaftliche Philosophie, 14*(3), 249–292. http://ontology.buffalo.edu/smith/book/FoGT/Ehrenfels_Gestalt.pdf.

Enriques, F., & Mazziotti, M. (1948). *Le dottrine di democrito d'abdera: testi e commenti.* Bologna: Nicola Zanichelli.

Enriques, F., & Mazziotti, M. (1982). Οι θεωρίες του Δημόκριτου του Αβδηρίτη - Κείμενα και Υπομνήματα. Ξάνθη: Έκδοση Διεθνούς Δημοκρίτειου Ιδρύματος.

Fairman, H. S., Brill, M. H., & Hemmendinger, H. (1997). How the cie 1931 color-matching functions were derived from wright-guild data. In *Color Research & Application: Endorsed by Inter-Society Color Council, The Colour Group (Great Britain), Canadian Society for Color, Color Science Association of Japan, Dutch Society for the Study of Color, The Swedish Colour Centre Foundation, Colour Society of Australia, Centre Français de la Couleur* (vol. 22(1), 11–23).

Frege, G. (1879). *Begriffsschrift, eine der arithmetischen nachgebildete formelsprache des reinen denkens.* Halle: Louis Nebert.

Frege, G. (1884). *Die grundlagen der arithmetik - eine logish mathematische untersuchung über den begriff der zahl.* Breslau: Wilhelm Koebner.

Frege, G. (1892). über sinn und bedeutung. *Zeitschrift für Philosophie und philosophische Kritik, 100*, 25–50.

Frege, G. (1893). *Grundgesetze der arithmetik - begriffsschriftlich abgeleitet* (Vol. I). Jena: Hermann Pohle.

Frege, G. (1918). *Der gedanke: Eine logische untersuchung.* Stenger.

Fresnel, A. (1866). *Oeuvres complètes d'augustin fresnel* (Vol. I). Paris: Imprimerie impériale.

Fresnel, A. (1868). *Oeuvres complètes d'augustin fresnel* (Vol. II). Paris: Imprimerie impériale.

Fresnel, A. (1870). *Oeuvres complètes d'augustin fresnel* (Vol. III). Paris: Imprimerie impériale.

Ganot, A. (1881). *Cours élémentaire de physique (Natural Philosophy, Translated and edited by E (Atkinson).* London: Longmans, Green and Co.

von Goethe, J. W.: (1810). *Zur farbenlehre.* https://theoryofcolor.org/Anzeige+und+Übersicht.

von Goethe, J. W. (1840). *Theory of colours (Translated from the German, with Notes by Charles Lock (Eastlake).* Albemarle Street, London: John Murray.

Grassmann, H. G. (1853). Zur theorie der farbenmischung. *Annalen der Physik, 165*(5), 69–84.

Grassmann, H. G. (1854). Theory of compound colours. *Philosophic Magazine, 4*(7), 254–264.

Guild, J. (1931). The colorimetric properties of the spectrum. *Philosophical Transactions of the Royal Society of London. Series A, Containing Papers of a Mathematical or Physical Character, 230*(681-693), 149–187. https://royalsocietypublishing.org/doi/10.1098/rsta.1932.0005.

Guild, J. (1970a). Interpretation of quantitative data in visual problems. In D. L. MacAdam (ed.), *Sources of colour science* (ISBN 0262130610 ed., pp. 213–245). Cambridge, Massachusetts and London, England: MIT Press.

Guild, J. (1970b). Some problems of visual perception." In D. L. MacAdam (ed.), *Sources of colour science* (ISBN 0262130610 ed., pp. 194–212). Cambridge, MA and London, England: MIT Press.

Hamilton, W. R. (1828a). Theory of systems of rays. *The Transactions of the Royal Irish Academy, 15*, 69–174. https://archive.org/details/jstor-30078906.

Hamilton, W. R. (1828b). "Theory of systems of rays." *The Transactions of the Royal Irish Academy*, *15*, 69–174. http://www.maths.tcd.ie/pub/HistMath/People/Hamilton/Rays.

Hamilton, W. R. (1830a). Second supplement to an essay on the theory of systems of rays. *The Transactions of the Royal Irish Academy*, *16*, 92–126. https://archive.org/details/jstor-30079028.

Hamilton, W. R. (1830b). Supplement to an essay on the theory of systems of rays. *The Transactions of the Royal Irish Academy*, *16*, 3–62. https://archive.org/details/jstor-30079022.

Hamilton, W. R. (1830c). Supplement to an essay on the theory of systems of rays. *The Transactions of the Royal Irish Academy*, *16*(1), 1–61. https://www.maths.tcd.ie/pub/HistMath/People/Hamilton/Rays.

Hamilton, W. R. (1831a). Second supplement to an essay on the theory of systems of rays. *The Transactions of the Royal Irish Academy*, *16*(2), 93–125. https://www.maths.tcd.ie/pub/HistMath/People/Hamilton/Rays.

Hamilton, W. R. (1831b). Third supplement to an essay on the theory of systems of rays. *The Transactions of the Royal Irish Academy*, *17*. https://archive.org/details/jstor-30078785.

Hamilton, W. R. (1837). Third supplement to an essay on the theory of systems of rays. *The Transactions of the Royal Irish Academy*, *17*(1), 1–144. https://www.maths.tcd.ie/pub/HistMath/People/Hamilton/Rays.

Heesen, R. (2015). *The young-(helmholtz)-maxwell theory of colour vision*. http://philsci-archive.pitt.edu/11279/.

von Helmholtz, H. L. F. (1852a). On the theory of compound colours. *The London, Edinburgh, and Dublin Philosophical Magazine and Journal of Science*, *4*(28), 519–534. https://bit.ly/3EZxeHZ.

von Helmholtz, H. L. F. (1852b). Ueber die theorie der zusammengesetzten farben. *Annalen der Physik*, *163*(9), 45–66. https://archive.org/details/bub_gb_UYAZAAAAYAAJ/mode/2up.

von Helmholtz, H. L. F. (1867). *Handbuch der physiologischen optik*. Leipzig: Leopold Voss.

von Helmholtz, H. L. F. (1885). The recent progress of the theory of vision [A cource of Lectures delivered in Frankfort and Heidelberg, and Republished in the Preussische Jahrbücher, 1868]. In E. Atkinson (ed.), *Popular lectures on scientific subjects* (Translated by E. Atkinson ed., pp. 197–316). 1, 3 and 5 Bond Street, New York: D. Appleton and Company. https://archive.org/details/popularlectureso00helmuoft.

von Helmholtz, H. L. F. (1909a). *Handbuch der physiologischen optik* (Third edn., Vol. I). Hamburg, Leipzig: Verlag von Leopold Voss.

von Helmholtz, H. L. F. (1909b). *Handbuch der physiologischen optik* (Third edn., Vol. II). Hamburg, Leipzig: Verlag von Leopold Voss.

von Helmholtz, H. L. F. (1909c). *Handbuch der physiologischen optik* (Third edn., Vol. III). Hamburg, Leipzig: Verlag von Leopold Voss.

Hering, E. (1868). *Die lehre vom binocularen sehen*. Leipzig: Von Wilhelm Engelmann.

Hering, E. (1879). Der raumsinn und die bewegungen des auges. In *Handbuch der physiologie* (Vol. III, pp. 434–601). Leipzig: F. C. W. Vogel. https://archive.org/details/handbuchderphys03unkngoog/mode/2up.

Hering, K. E. K. (1878). Principles of a new theory of the color sense. In R. C. Teevan & R. C. Birney (eds.), *Color vision: An enduring problem in psychology* (English translation by K. Butler, 1961 ed.). New York: Van Nostrand.

Hering, K. E. K. (1899). über die grenzen der sehschärfe. *Berichte über die Verhandlungen der Koniglich-Sächsischen Gesellschaft der Wissenschaften zu Leipzig. Mathematisch-Physische Klasse*, *51*, 16–24. http://hans-strasburger.userweb.mwn.de/materials/hering_1899.pdf.

Hering, K. E. K. (1920). *Grundzüge der lehre vom lichtsinn*. Berlin: Von Julius Springer.

Hering, K. E. K. (1964). *Outlines of a theory of the light sense* (Translated by Leo M. Hurvich and Dorothea Jameson ed.; L. M. Hurvich & D. Jameson, eds.). Cambridge, MA: Harvard University Press.

Hett, W. S. (1935). *Aristotle on the soul*. London, UK: Willian Heinemann Ltd.

Huygens, C. (1690). *Traité de la lumière*. chez Pierre vander Aa, Marchand Libraire.

Huygens, C. (1912). *Treatise on light* (English translation by Silvanus P. Thomson ed.). St. Matrin's Street, London: Macmillan and Co., Limited.

Huygens, C., Young, T.., & Fresnel, A. J. (1900). In H. Crew (ed.), *The wave theory of light: Memoirs of huygens, young and fresnel*. New York: American Book Company.

Ierodiakonou, K. (2005). Empedocles on colour and colour vision. *Oxford Studies in Ancient Philosophy, 29*(1), 38.

Isaacson, W. (2017). *Leonardo da vinci*. Μετάφραση στα Ελληνικά Γιώργος Μπαρουξής, Εκδόσεις Ψυχογιός, 2018. New York: Simon & Schuster.

Kant, I., & (1787). Critique of pure reason, 1900. *Second Edition Revised, Translated by J. M. D* (Meiklejohn). New York: P. F. Collier and Son.

Keele, K. D. (1955). Leonardo da vinci on vision. *Proceedings of the Royal Society of Medicine, 48*(5), 384–390. https://journals.sagepub.com/doi/pdf/10.1177/003591575504800512.

Kepler, J. (1604). *Optics, paralipomena to witelo and the optical part of astronomy*. Santa Fe: Green Lion Press (English translation by William H. Donahue).

Kepler, J. (1611). *Dioptrice*. Augustae Vindelicorum, typis Favidis Franci.

Kepler, J., & Donahue, W. H. (2000). *Optics: Paralipomena to witelo & optical part of astronomy* (Translated by William H. Donahue, ISBN 1-888009-12-8 ed.). Santa Fe, New Mexico: Green Lion Press.

Koffka, K. (1924). *The growth of the mind: An introduction to child-psychology* (Translated by Robert Morris Ogden ed.; R. M. Ogden, ed.). New York: Harcourt, Brace.

Koffka, K. (1935). *Principles of gestalt psychology*. New York: Harcourt, Brace and Company.

Köhler, W. (1847). *Gestalt psychology: An introduction to new concepts in modern psychology* (Renewed 1975 by Lilli Köhler, Reissued 1992. ISBN 0-87140-218-1 ed.). 500 Fifth Avenue, New York: Liveright Publishing Corporation.

Köhler, W. (1925). *The mentality of apes* (1956, Translated from the Second Revised Edition by Ella Winter B.Sc. ed.; E. Winter, ed.). Broadway House 68–74 Carter Lane, London: Routledge & Kegan Paul Ltd.

König, A., & Dieterici, C. (1886). Die grundempfindungen und ihre intensitäts-vertheilung im spectrum. *Sitzungsberichte der Akademie der Wissenschaften zu Berlin*, 805–829.

König, A., & Dieterici, C. (1892). *Die grundempfindungen in normalen und anomalen farbensystemen und ihre intensitätsverteilung im spektrum*. Hamburg, Leipzig: Leopold Voss.

Kremer, R. L. (1993). Innovation through synthesis: Helmholtz and color research. *Hermann von Helmholtz and the foundations of nineteenth-century science*, 205–258.

von Kries, J. (1878). Beitrag zur physiologie der gesichtsempfindungen. In W. His, W. Braune., & E. D. Bois-Reymond (eds.), *Archiv für anatomie und physiologie* (Vol. I und II Heft, pp. 503–524). Leipzig: Veit & Comp. https://www.europeana.eu/en/item/368/item_OOO2Z6MB4IRBCRAFOL2DSO7K3W6LP6N7.

von Kries, J. (1905). Die gesichtsempfindungen. In W. A. Nagel (ed.), *Handbuch der physiologie des menschen* (Vols. Dritter Band, Physiologieder Sinne, pp. 109–282). Braunschweig: Druck und Verlag von Friedrich Vieweg und Sohn. https://archive.org/details/handbuchderphysi03nage.

Lee, H. D. P. (1952). *Aristotle meteorologica*. London, UK: Willian Heinemann Ltd.

Llinás, R. R. (2003). The contribution of Santiago Ramon Y Cajal to functional neuroscience. *Nature Reviews Neuroscience, 4*(1), 77–80.

Longair, M. S. (2008). Maxwell and the science of colour. *Philosophical Transactions of the Royal Society A: Mathematical, Physical and Engineering Sciences, 366*(1871), 1685–1696.

MacAdam, D. L. (1970). *Sources of colour science*. Cambridge, MA and London, England: The MIT Press. 0-262-13061-0

MacCurdy, E. (1955). *The notebooks of leonardo da vinci*. New York: George Braziller.

Maxwell, J. C. (1855). On the theory of colours in relation to colour-blindness. *Transactions of the Royal Scottish Society of Arts, IV*(III). https://archive.org/details/scientificpapers01maxwuoft/mode/2up.

Maxwell, J. C. (1856a). On the theory of compound colours with reference to mixtures of blue and yellow light. *Report of the British Association*, (2), 12–13. https://archive.org/details/scientificpapers01maxwuoft/mode/2up.

Maxwell, J. C. (1856b). On the unequal sensibility of the foramen centrale to light of different colours. *Report of the British Association*. https://archive.org/details/scientificpapers01maxwuoft/mode/2up.

Maxwell, J. C. (1856). Theory of the perception of colours. *Transactions of the Royal Scottish Society of Arts, 4*, 394–400.

Maxwell, J. C. (1857a). An account of experiments on the perception of colour. *The London, Edinburgh, and Dublin Philosophical Magazine and Journal of Science, XIV*(90), 40–47.

Maxwell, J. C. (1857). The diagram of colours. *Transactions of the Royal Society of Edinburgh, 21*, 275–298.

Maxwell, J. C. (1857c). Experiments on colour, as perceived by the eye, with remarks on colourblindness. *Transactions of the Royal Society of Edinburgh, XXI*(II), 275–298.

Maxwell, J. C. (1858). On the general laws of optical instruments. *The Quarterly Journal of Pure and Applied Mathematics, 2*, 233–246.

Maxwell, J. C. (1860). On the theory of compound colours, and the relations of the colours of the spectrum. *Philosophical Transactions of the Royal Society of London, 150*, 57–84. http://www.jstor.org/stable/108759.

Maxwell, J. C. (1861). *On the theory of three primary colours*. Lecture at the Royal Institution of Great Britain, May 17, 1861. https://archive.org/details/scientificpapers01maxwuoft/mode/2up.

Maxwell, J. C. (1867). On the cyclide. *The Quarterly Journal of Pure and Applied Mathematics*, (34), 111–126. https://archive.org/details/scientificpapers02maxwuoft/mode/2up.

Maxwell, J. C. (1869). On the best arrangement for producing a pure spectrum on a screen. *Proceedings of the Royal Society of Edinburgh, 6*, 238–242.

Maxwell, J. C. (1871a). On colour vision. *Nature, 4*, 13-16. https://www.nature.com/articles/004013b0.

Maxwell, J. C. (1871). On the focal lines of a refracted pencil. *Proceedings of the London Mathematical Society, 1*(1), 337–343.

Maxwell, J. C. (1872). On colour vision. *Proceedings of the Royal Institution of Great Britain, VI*, 260–271.

Maxwell, J. C. (1874). On hamilton's characteristic function for a narrow beam of light. *Proceedings of the London Mathematical Society, 1*(1), 182–190. https://archive.org/details/scientificpapers02maxwuoft/mode/2up.

McCarty, D. C., et al. (2000). Optics of thought: Logic and vision in müller, helmholtz, and frege. *Notre Dame Journal of Formal Logic, 41*(4), 365–378.

Müller, J. (1838). *Handbuch der physiologie des menschen für vorlesungen (Vol. I and II)*: Goblenzm Verlag von J. Hölscher.

Müller, J. P. (1835). *Handbuch der physiologie des menschen für vorlesungen* (Vol. I). Coblenz: J. Hölscher.

Müller, J. P. (1840a). *Elements of physiology* (Translated from the German with notes by Willian Bally, M.D. ed., Vol. I). Upper Gower Street, London: Taylor and Walton.

Müller, J. P. (1840). *Handbuch der physiologie des menschen für vorlesungen* (Vol. II). Coblenz: J. Hölscher.

Müller, J. P. (1842). *Elements of physiology* (Translated from the German with notes by Willian Bally, M.D. ed., Vol. II). Upper Gower Street, London: Taylor and Walton.

Newton, I. (1671). New theory about light and colours. *Philosophical Transactions of the Royal Society of London, 80*, 3075–3087.

Newton, I. (1687). *Philosophiæ naturalis principia mathematica*. London: Jussu Societatis Regiae AC Typis Josephi Streater. Prostat apud plures bibliopolas.

Newton, I. (1704). *Opticks: Or, a treatise of the reflexions, refractions, inflexions and colours of light*. Printed for Sam. Smith. and Benj. Walford, Printers to the Royal Society, at the Prince's Arms of St. Pauls Curch-yard, London.

Niall, K. K. (2017). Erwin schrödinger's color theory: Translated with modern commentary. In K. K. Niall (Ed.), *Leonardo*.

Palmer, G. (1777). *Theory of colours and vision*. London: Leacroft.

Palmer, G. (1786). *Theorie de la lumiere, applicable aux arts, et principalement a la peinture*. Paris: Hardouin et Gattey. http://gallica.bnf.fr/ark:/12148/bpt6k74566p/f1.item

Picard, J. (1671). *Mesure de la terre*. Paris: De Limprimerie Royale.

Plato. (2008). *Timaeus*. Ekdosis Polis (Original work published ca.370 BCE). ISBN: 9789607478115

Plato. (2014). *The republic (politeia)*. Ekdosis Polis (Original work published ca.380 BCE). ISBN: 9789608132719

Polyak, S. L. (1941). *The retina: The anatomy and the histology of the retina in man, ape, and monkey, including the consideration of visual functions, the history of physiological optics, and the histological laboratory technique*. Chicago: University of Chicago Press.

Polyak, S. L. (1949). Retinal structure and colour vision. *Documenta Ophthalmologica, 3*, 24–56. https://link.springer.com/article/10.1007/BF00162597.

Polyak, S. L. (1957). *The vertebrate visual system; its origin, structure, and function and its manifestations in disease with an analysis of its role in the life of animals and in the origin of man, preceded by a historical review of investigations of the eye, and of the visual pathways and centers of the brain*. Chicago: University of Chicago Press.

Polyak, S. L. (1970). Retinal structure and color vision. In D. L. MacAdam (Ed.), *Sources of colour science* (pp. 246–268). Cambridge, Massachusetts and London, England: MIT press.

Ramón y Cajal, S. (1894a). The croonian lecture–la fine structure des centres nerveux. *Proceedings of the Royal Society of London, 55*(331–335), 444–468.

Ramón y Cajal, S. (1894b). *Les nouvelles idées sur la structure du système nerveux chez l'homme et chez les vertébrés* (French edition, Translated by Léon Azoulay ed.). Rue des Saints-Pères 15, Paris: C. Reinwald & Cie.

Ramón y Cajal, S. (1899). *Textura del sistema nervioso del hombre y de los vertebrados: estudios sobre el plan estructural y composición histológica de los centros nerviosos adicionados de consideraciones fisiológicas fundadas en los nuevos descubrimientos* (Vol. 1). Carretas 8 y Garcilaso 6, Mardid: Nicolás Moya.

Ramón y Cajal, S. (1909). *Histologie du système nerveux de l'homme & des vertébrés* (French edition, Translated by Léon Azoulay ed., Vol. 1). Rue De l'École-de-Médecine 25–27, Paris: A. Maloine.

Ramón y Cajal, S. (1911). *Histologie du système nerveux de l'homme & des vertébrés* (French edition, Translated by L. Azoulay ed., Vol. 2). Rue De l'École-de-Médecine 25–27, Paris: A. Maloine.

Risnero, F. (1572). *Opticae thesaurus*. Basileae, Per Episcopius.

Runge, P. O. (1810). *Farben-kugel*. Fredrick Perthes.

Schopenhauer, A. (1816). *Ueber das sehn und die farben eine abhandlung*. Johann Friedrich Hartknoch.

Schopenhauer, A. (1969a). *The world as will and representation* (1958 Edition, Translated by E. F. J. Payne ed., Vol. I). New York: Dover Publications, Inc.

Schopenhauer, A. (1969b). *The world as will and representation* (1958 Edition, Translated by E. F. J. Payne ed., Vol. II). New York: Dover Publications, Inc.

Schopenhauer, A., & Runge, P. O. (2010). *On vision and colors & color sphere* (Translated and with an introduction by Georg Stahl, ISBN 978-1-61689-005-6 ed.). 37 East Seventh Street, New York, New York 10003: Princeton Architectural Press.

Schrödinger, E. (1920a). Grundlinien einer theorie der farbenmetrik im tagessehen, part 1. *Annalen der Physik, 368*(21), 397–426. https://onlinelibrary.wiley.com/doi/10.1002/andp.19203682102.

Schrödinger, E. (1920b). Grundlinien einer theorie der farbenmetrik im tagessehen, part 2. *Annalen der Physik, 368*(21), 427–456. https://onlinelibrary.wiley.com/doi/10.1002/andp.19203682103.

Schrödinger, E. (1920c). Grundlinien einer theorie der farbenmetrik im tagessehen, part 3. *Annalen der Physik, 368*(21), 481–520. https://onlinelibrary.wiley.com/doi/pdf/10.1002/andp.19203682202.

Schrödinger, E. (1925). über das verhältnis der vierfarben- zur dreifarbentheorie. *Sitzungberichte. Abt. 2a, Mathematik, Astronomie, Physik, Meteorologie und Mechanik. Akademie der Wissenschaften in Wien, Mathematisch-Naturwissenschaftliche Klasse, 134*, 471–490.

Schrödinger, E. (1994). On the relationship of four-color theory to three-color theory. *Color Research and Application, 19*(1), 37–47.

Sedley, D. N. (1992). *Empedocles' theory of vision and theophrastus' de sensibus* (pp. 20–31). Theophrastus: His Psychological, Doxographical and Scientific Writings.

Service, P. (2016). *The wright-guild experiments and the development of the cie 1931 rgb and xyz color spaces*. https://philservice.typepad.com/Wright-Guild_and_CIE_RGB_and_XYZ.pages. pdf.

Sherman, P. D. (1981). *Color vision in the nineteenth century: The young-helmholtz-maxwell theory*. Bristol: Adam Hilger Ltd. (ISBN 978-0852743768 ed.)

Shorey, P. (1935a). *The republic* (Vol. I (Books I-V)). Willian Heinemann Ltd.

Shorey, P. (1935b). *The republic* (Vol. II (Books VI-X)). Willian Heinemann Ltd.

Smith, T., & Guild, J. (1931). The cie colorimetric standards and their use. *Transactions of the optical society, 33*(3), 5–134. https://iopscience.iop.org/article/10.1088/1475-4878/33/3/301/pdf?casa_token=ZOBPYZ3BTR0AAAAA:JLHBrDEzLDKXN8X9p6r-gOfI8qBZBh0ujO9v5iGKHbtK6H7pF3j8x8WyeaST0Bn0qM9E6zWZiOk.

Stockman, A., & Sharpe, L. T. (2000). The spectral sensitivities of the middle-and long-wavelength-sensitive cones derived from measurements in observers of known genotype. *Vision Research, 40*(13), 1711–1737.

Strasburger, H., Huber, J., & Rose, D. (2018). Ewald hering's (1899) on the limits of visual acuity: A translation and commentary: With a supplement on alfred volkmann's (1863) physiological investigations in the field of optics. *i-Perception, 9*(3), 1–14. https://journals.sagepub.com/doi/full/10.1177/2041669518763675.

Thomas, T. (1808). *The treatises of aristotle on the soul*. London, UK: Taylor Thomas, by Robert Wilks.

Tredennick, H. (1933). *Aristotle: The metaphysics, books I–IX*. London, UK: Willian Heinemann Ltd.

Triarhou, L. C. (2007). Stjepan Poljak (1889–1955). *Journal of Neurology, 254*(11), 1619.

Turner, R. S. (1994). Consensus and controversy: Helmholtz on the visual perception of space. In D. Cahan (ed.), *Hermann von Helmholtz and the foundations of nineteenth-century science* (ISBN: 0520083342 ed., pp. 154–204). Berkeley and Los Angeles, CA: University of California Press.

Wachtler, J. (1896). *De alcmaeone crotoniata*. Teubneri: Lipsiae In aedibus B.G.

Wertheimer, M. (1912). Experimentelle studien über das sehen von bewegung. *Zeitschrift für Psychologie, 61*(1), 162–265. http://gestalttheory.net/download/Wertheimer1912_Sehen_von_Bewegung.pdf.

Wertheimer, M. (1922). Untersuchungen zur lehre von der gestalt i. *Psychologische Forschung, 1*, 47–58. https://doi.org/10.1007/BF00410385.

Wertheimer, M. (1923). Untersuchungen zur lehre von der gestalt ii. *Psychologische Forschung, 4*, 301–350. https://doi.org/10.1007/BF00410640.

Wittgenstein, L. (1922). *Tractatus logico-philosophicus (1922 Edition with an Introduction by Bertrand (Russell)*. London: Routledge & Kegan Paul Ltd.

Wittgenstein, L. (1977). *Bemerkungen über die farben (Translated by McAlister*. Margarete: Linda L. and Schättle. , 9780520033351 ed.; G. E. M. Anscombe, ed.). Berkeley and Los Angeles, CA: University of California Press.

Wold, J. H., & Valberg, A. (1999). *Mathematical description of a method for deriving an xyz tristimulus space*. Department of Physics: University of Oslo.

Wright, W. D. (1929). A re-determination of the trichromatic coefficients of the spectral colours. *Transactions of the Optical Society, 30*(4), 141.

Young, T. (1802). The bakerian lecture: On the theory of light and colours. *Philosophical Transactions of the Royal Society of London*, *92*, 12–48. https://archive.org/details/jstor-107113/page/n1/mode/2up.

Young, T. (1807a). *A course of lectures on natural philosophy and the mechanical arts* (Vol. I). Bedford Bury, London: For Joseph Johnson by William Savage.

Young, T. (1807b). *A course of lectures on natural philosophy and the mechanical arts* (Vol. II). Bedford Bury, London: For Joseph Johnson by William Savage.

Young, T. (1807c). Lecture xl lecture xl on the history of optics. In *A course of lectures on natural philosophy and the mechanical arts* (Vol. I, pp. 472–482). Bedford Bury, London: For Joseph Johnson by William Savage. https://archive.org/details/lecturescourseof01younrich.

Zajonc, A. G. (1976). Goethe's theory of color and scientific intuition. *American Journal of Physics, 44*(4), 327–333.

Printed in the United States
by Baker & Taylor Publisher Services